Blockchain and Artificial Intelligence-Based Solution to Enhance the Privacy in Digital Identity and IoT

The chapters of this book explore the main domains that represent considerable risks for the respect of privacy, such as education, health, finance, or social media.

Through its place in the massive data production industry, the Internet of Things participates in the development of artificial intelligence and is increasingly attracting the attention of web giants, governments, and especially all types of hackers. Thanks to this book, private and public organizations will have at their disposal a tool that highlights, on the one hand, the major challenges raised by privacy in the context of the Internet of Things and, on the other hand, recommendations for improving good practices.

Digital identity is presented as a bulwark for the protection of privacy. It opens up new avenues for improving digital trust. Concretely, there are a set of challenges that are associated with the management of digital identity, mainly in relation to the compliance and governance of personnel data in order to eliminate privacy and security risks.

Blockchain and Artificial Intelligence-Based Solution to Enhance the Privacy in Digital Identity and IoT

Edited by
Fehmi Jaafar and Schallum Pierre

CRC Press is an imprint of the
Taylor & Francis Group, an **informa** business

First edition published 2024
by CRC Press
2385 Executive Center Drive, Suite 320, Boca Raton, FL 33431

and by CRC Press
4 Park Square, Milton Park, Abingdon, Oxon, OX14 4RN

CRC Press is an imprint of Taylor & Francis Group, LLC

ISBN: 9781032130989 (hbk)
ISBN: 9781032131009 (pbk)
ISBN: 9781003227656 (ebk)

DOI: 10.1201/9781003227656

Typeset in Times
by KnowledgeWorks Global Ltd.

Contents

About the Authors vii
List of Contributors ix

Introduction 1
Schallum Pierre and Fehmi Jaafar

SECTION I Digital Identity Era

Chapter 1 Demystifying the Digital Identity Challenges and the Blockchain Role 9
Manel Grichi and Fehmi Jaafar

Chapter 2 Blockchain for Digital Identity 25
Tim Weingärtner

SECTION II Privacy Dilemma

Chapter 3 Security and Corporate Violation to Privacy in the Internet of Things Age 39
Darine Amayed, Fehmi Jaafar, Riadh Ben Chaabene and Mohamed Cheriet

Chapter 4 Security, Privacy, and Blockchain in Financial Technology 71
Schallum Pierre and Olson Italis

SECTION III Sensitive Data Challenges

Chapter 5 Where Does the Novel Legal Framework for AI in Canada Stand against the Emerging Trend of Online Test Proctoring? 97
Céline Castets-Renard and Simon Robichaud-Durand

Chapter 6 Blockchain, AI, and Data Protection in Healthcare: A Comparative Analysis of Two Blockchain Data Marketplaces in Relation to Fair Data Processing and the 'Data Double-Spending' Problem 125

Deepansha Chhabra, Meng Kang and Victoria Lemieux

Chapter 7 Cyber Influence Stakes 155

Rachel Ladouceur and Fehmi Jaafar

Postface 175

Schallum Pierre and Fehmi Jaafar

Index 179

About the Authors

Dr. Fehmi Jaafar is currently an associate professor at Québec University at Chicoutimi and an affiliate professor at Laval University and Concordia University.

Previously, he was a researcher at the Computer Research Institute of Montréal, an adjunct professor at Concordia University of Edmonton, and a postdoctoral research fellow at Queen's University and Polytechnique Montréal.

Dr. Fehmi Jaafar received his PhD from the Department of Computer Science at Montréal University, Canada. He is interested in cybersecurity in the Internet of Things, in the analysis and improvement of the security and quality of software systems, and in the application of machine learning techniques in cybersecurity.

His research has been published in top venues in computer sciences, including the *Journal of Empirical Software Engineering* (EMSE) and the *Journal of Software: Evolution and Process* (JSEP). He established externally funded research programs in collaboration with Defence Canada, Safety Canada, Natural Sciences and Engineering Research Council of Canada (NSERC), Mathematics of Information Technology and Complex Systems (MITACS), industrial partners, and foreign universities.

Dr. Schallum Pierre is an EDI, Ethics and Innovation advisor within Institute of Intelligence and Data (IID) at Université Laval and a part-time professor in social communications at Saint Paul University. He contributes to the implementation of ethics by design in AI research projects and in data valorization at the IID. As an innovation advisor, he accompanies public and private organizations in R&D projects and in the National Research Council of Canada Industrial Research Assistance Program (NRC-IRAP) at Université Laval. His rich and varied experience in responsible entrepreneurship is put to use in collaborations with the socio-economic milieu of the greater Québec City area in knowledge and technology transfer related to AI and data valorization.

As a researcher, Schallum is interested in ethical issues related to AI and Blockchain. His research interests include the issue of protecting sensitive data in the healthcare, social media, and mobile payment sectors. He holds a PhD in philosophy from Université Laval, and he completed a postdoctoral fellowship at Polytechnique Montréal.

He is a member of the International Association of Privacy Professionals, the Ethics Committee of the CHU de Québec – Université Laval, the Open Government Committee of Québec, and the editorial board of the journal *Technologie et innovation.*

Contributors

Darine Ameyed received a PhD in Engineering, Software, and Information Technologies from ÉTS, Canada, in 2017. She received a certificate in applied IA from Columbia University in 2019. She obtained her MSc in Digital Art and Technology from the University Rennes 2 (UR2) (University of Upper Brittany, France) in 2010. She earned her MSc Multimedia Engineering from the University Paris Est Marne la Vallée (UPEM) (University of Paris 11, France) in 2008. She received a BSc in Computer Science and Management from the Institut Supérieure de Gestion (ISG) (University of Tunis, Tunisia) in 2005. She is currently a researcher and AI advisor at the government of Québec, associate researcher at Synchromedia lab, École de Technologie Supérieure (ÉTS), Montréal, and affiliate professor at University of Sherbrooke. She is Tech-savvy and research-driven professional with comprehensive experience conducting research in prediction modeling, Context-Aware System, human activity recognition, Cognitive IoT, Machine Learning, and Ambient Intelligence. Her research interests include Human-Centered Computing, Human–Machine Interaction (HMI), Data security and privacy in IoT platforms and computational sustainability. On the other hand, she has a long multidisciplinary academic and industrial career in Canada, Europe, and Africa in the areas of software engineering, mobile computing, ERP system management, and entrepreneurship. She is co-founder and chief scientist officer at Baüne. She was, also, an expert in United for Smart and Sustainable Cities (U4SSC)-International Telecommunication Union-ITU-United Nation. She has been a scientific program and project manager at CIRODD (2015–2019), co-founder and CEO of NyX-R (2015–2018), and a tech advisor for tech and science-based startups.

Riadh Ben Chaabene is a research master's student from the Software and Information Technology Engineering Department at École de technologie supérieure ÉTS Montréal, Québec, Canada. He has a bachelor's degree (Licence degree) in business computing from the University of Tunisia in 2018 and a Master's degree in Artificial Intelligence and Blockchain from the Mediterranean Institute of Technology in 2021. His current research interest includes machine learning, federated learning, privacy preservation, and cybersecurity.

Céline Castets-Renard is the research chairholder accountable AI in a Global Context. She also holds a research chair in Law, Accountability, and Social Trust in AI at ANITI (Artificial and Natural Intelligence Toulouse Institute) in France. She is a member expert at the European Commission, Observatory Online Platform Economy. She is a former junior member at the "Institut Universitaire de France" and Fellow at the Internet Society Project, Law Faculty, Yale University. Her research generally focuses on the law and regulation of digital technologies and artificial intelligence in a comparative perspective (Canadian, European, and American law), especially the protection of personal data and privacy, policing technologies, online platforms, and cybersecurity. She also studies the impact of technologies on human rights, equity, and social justice in a global perspective, particularly in north-south relationships.

Mohamed Cheriet is a senior member of IEEE. He received M.Sc. and PhD degrees in computer science from the University of Pierre and Marie Curie (Paris VI) in 1985 and 1988, respectively. Since 1992, he has been a professor with the Systems Engineering Department, École de Technologie Supérieure (ETS), University of Québec, Montréal, where he was appointed as a full professor in 1998. Since 1998, he has been the founder and the director of the Synchromedia Laboratory for multimedia communication in telepresence applications. He is currently an expert in computational intelligence, pattern recognition, machine learning, artificial intelligence, and perception. He published six patents (three granted), and the first standard ever, IEEE 1922.2, on a real-time calculation of Information and Communications Technology (ICT) emissions, in April 2020, with the IEEE Emissions Working Group. He has published more than 500 technical articles in the field and serves on the editorial boards of several renowned journals and international conferences. His research has extensive experience in sustainable and empowered AI future networks and cloud computing. He is also the administrative director of the CFI'2020 CEOS Network. He was a fellow of the International Association of Pattern Recognition (IAPR) in 2016, the Canadian Academy of Engineering (CAE) in 2017, the Engineering Institute of Canada (EIC) in 2018, and the Engineers Canada (EC) in 2019. He is a Steering Committee member of the IEEE Sustainable ICT Initiative. He was a recipient of the 2016 IEEE J.M. Ham Outstanding Engineering Educator Award; the ÉTS Research Excellence Prize in 2013, for his outstanding contribution to green ICT, cloud computing, and big data analytics research areas; and the 2012 Queen Elizabeth II Diamond Jubilee Medal. He held a 2013 Tier 1 Canada Research chair on sustainable and smart eco-cloud and led the establishment of the first smart-university campus in Canada, created as a hub for innovation and productivity at Montréal. He is also the general director of the FRQNT Strategic Cluster on the operationalization of sustainability development, CIRODD (2019–2026). He is also the founder and former chair of the IEEE Montréal Chapter of Computational Intelligent Systems (CIS) and the chair of the ICT Emissions Working Group.

Deepansha Chhabra is a Master of Engineering graduate from the Department of Electrical and Computer Engineering at the University of British Columbia. She completed her course-based Master's focused on Blockchain. During the course of her study at University of British Columbia (UBC), she worked as a Graduate research assistant under the supervision of Dr. Victoria Lemieux. Her research was based on analyzing the fairness of data processing, preventing the unauthorized secondary usage of data, and the double-spending problem in the Ethereum-based Ocean Protocol.

Manel Grichi received a Computer Engineering PhD in 2020 from the Department of Computer Science and Software Engineering at Polytechnique Montréal, Québec, Canada. She has a bachelor's degree (Licence degree) and a software engineering's degree from the University of Tunisia in 2011 and 2014, respectively. Her current research interests include mining software repositories, data analysis, machine learning, and software quality/security.

Olson Italis is currently a PhD candidate in Computer Engineering, in a bi-diploma program at the Institute of Science, Technology and Advanced Studies of Haiti (ISTEAH) and at Polytechnique Montréal. Attached to the Networking and Mobile Computing Research Laboratory (LARIM), his field of research includes computer security, electronic commerce, and Blockchain. He is a lecturer at ISTEAH since fall 2020 and a lecturer at Polytechnique since winter 2022. He is also actively involved in the student associations of ISTEAH (AEE-ISTEAH) and Polytechnique (AÉCSP). Wanting to give back what he has obtained, he participated in Polytechnique's "peer mentoring" program to help new students in their careers. Mr. Italis graduated from the Faculty of Sciences (FDS) of the State University of Haiti (UEH) with a degree in Electronic Engineering and then went on to earn a Master's degree in Computer Engineering and Information Technology from ISTEAH. Meanwhile, he worked at the Bank of the Republic of Haiti as a programmer analyst from 2008 to 2019. He was a teaching assistant in mathematics at the propedeutic studies of the FDS (2006–2009), in charge of the Introduction to Computer Science course at the Centre Technique de Planification et d'Économie Appliquée (CTPEA), from 2009 to 2017.

Meng Kang is currently a research engineer at Lifeguard Digital Health. He received a Master of Applied Science degree in Electrical Engineering from the University of British Columbia. Currently, he is working on a blockchain-enabled secure and privacy-preserving solution and machine learning for IoT wearables in digital health. Previously, he conducted research on decentralized identity-based blockchain for the prevention of unauthorized secondary data usage under the supervision of Dr. Victoria Lemieux.

Rachel Ladouceur, CPA, CISA, is a PhD student in Information Science and Technology at the Université du Québec à Chicoutimi (UQAC), Canada. She also holds a Master's degree in management from HEC Montréal and a Master's degree in information technology from ETS. She worked for 13 years at the Government of Québec in information security and 7 years at the Auditor General of Québec as an expert in computer auditing. She was in charge of IT performance audit mandates. She is interested in the detection of disinformation on social networks and artificial intelligence, specifically NLP.

Victoria Lemieux is a professor of Archival Science at the University of British Columbia (UBC) School of Information. She is also founder of Blockchain@UBC, the University of British Columbia's Blockchain research and education cluster, education director of UBC's blockchain graduate pathway, a Sauder School of Business Distinguished Scholar, and a faculty associate of the Institute for Computing, Information and Cognitive Systems at UBC. Her latest publications include *Building Decentralized Trust: Multidisciplinary Perspectives on the Design of Blockchains and Distributed Ledgers* (Springer, 2021) and *Searching for Trust: Blockchain in an Age of Disinformation* (Cambridge University Press, 2022).

Simon Robichaud-Durand is a Masters of Law (LLM) and Concentration Law and Technology student at the University of Ottawa, who works on artificial intelligence, privacy, and criminal law topics. He holds a Licentiate in Law (LLL) and an Honours BSocSc International Development and Globalization degree from the University of Ottawa, where he focused his studies on technological legal issues, such as the legal framework for cyberattacks, and environmental and globalization issues. Currently, he works for the Accountable Artificial Intelligence in a Global Context research chair where he has been involved in the creation of a report on the use of proctoring software by Canadian universities for the Office of the Privacy Commissioner of Canada, as well as other French and English publications on the topic. For his Masters studies, Simon's research is focused on the intersection of artificial intelligence, social media networks, and the protection of minors.

Dr. Tim Weingärtner is a professor at the School of Computer Science and Information Technology at the Lucerne University of Applied Sciences and Arts, Switzerland. In lecturing and research, he focuses on blockchain technology, DLT, and Smart Contracts. As a representative in the Smart-up Program, he supports the promotion of young startups from the university. He is president of DEC – DLT Education Consortium, expert member of the EU Blockchain Observatory and Forum, and board member of DIDAS – Digital Identity and Data Sovereignty Association, Switzerland.

Introduction

Schallum Pierre
Institut intelligence et données (IID),
Université Laval, Québec, Canada

Fehmi Jaafar
Department of Mathematics and Computer Science,
Québec University at Chicoutimi, Québec, Canada

CONTENTS

Our Sensitive Data 2
Our Private Life 3
Sensitive Data Raises Privacy Issues 4

Our contemporary societies are producing more and more data. Between January 1, 2022, and August 20, 2022 alone, 580,294,915,326 Gigabytes (GB) of data were published, according to Planetoscope.[1] As reported by the same source, we produce, around the world, an average of 29,000 (GB) of data, every second. That's 2.5 exabytes per day or 912.5 exabytes per year. Among these data, IoT is an important source. In 2022, the number of IoT devices is estimated to be over 13.1 billion.[2] By 2030, this figure will rise to 29.4 billion. Exploiting the data generated in the context of sensors, IoT and industrial environments remains a great challenge. It is estimated that today only 20% of industrial data has already been used.[3] This amount of data needs to be valorized to be put at the service of research, citizens, and the advancement of the understanding of climate issues. This is why data plays a key role in digital transformation. It must be collected, stored, and processed. Because of their ability to deal with targeted and personalized information, IoT has become very present in our environment. Its place in our daily lives explains this scope: from the moment we wake up until late at night. For example, during the pandemic, different segments of the world's population, especially the elderly, were confronted with sleep problems. Faced with this need, IoT technologies have been used in monitoring the quality of sleep,[4] without the need for a person to go to the hospital to expose himself to risks of contagion. Various categories of personnel data are generated by IoT: health data, geolocation data, financial data, sociodemographic data, biometric data, etc. Several standards govern IoT that are the source of unstructured data such as movement, gesture recognition, and biometric measurement, proximity; infrared luminosity; sound, and; smell. These standards include IEEE 802.11.[5] Two technologies are helping to leverage this data and accelerate digital transformation: artificial intelligence (AI) and blockchain.

DOI: 10.1201/9781003227656-1

In AI, particularly supervised learning, historical data obtained from IoT can be used to train models that will be used to automate useful tasks in a human environment. A house with video-based fire detection systems will be mapped. All the data generated will allow the AI to act like humans in that space. The people living there will be able to be warned of possible heat, smoke, and fire hazards.[6] There are several security risks[7] that can make IoT vulnerable. Attacks can exploit known vulnerabilities and backdoors[8] in IoT devices. Backdoors create problems with confidentiality, integrity, and authentication. As an example, a backdoor in a pacemaker[9] can put a patient's life at risk. Certainly, the collection of cardiac event data, used by AI, can help the patient make better decisions, but if that data as well as access to the device falls into malicious hands, it can cause irreparable harm.

The collection of data, which is essential for AI, can be both useful and dangerous. The problem in Afghanistan with the Taliban's takeover of biometric databases[10] is worthy of our consideration. Biometric information of thousands of Afghans collected by the US military was abandoned. As Intercept recalls, this database includes data such as iris scans and fingerprints.[11] Since this data is not protected, it is used against opponents of the regime. This extreme example raises the issue of personal data protection.

Should we run the risk of giving a centralizing organization such as a private company or a government control over our personal data? What are the issues related to the collection, storage, and use of personal data? What role can blockchain play in a governance that focuses on privacy protection? First, why should we pay so much attention to the governance of the data we produce?

OUR SENSITIVE DATA

According to the General Data Protection Regulation (GDPR),[12] when the processing of a personal data can represent a significant risk to fundamental rights and freedoms, it is sensitive data. Consequently, all security measures must be considered. Without limitation, the sensitive nature may concern data related to:

- ethnicity
- fingerprints
- biometrics
- financial
- education
- sexual preference
- social media
- expression of opinion (political, religious, and union)
- vulnerable persons
- minors

The collection of data by IoT can concern not only sensitive personal data but also another category of sensitive data[13] which are industrial data or business data.

These categories of data (such as industrial data and sensitive personal data) must be protected, at the risk of revealing confidential information about an organization or a person. In the case of an organization, it is the sharing of strategic data. In 2021, Canada took steps to protect the research[14] sector from possible threats or espionage. In the case of an individual, it is the sharing of privacy.

OUR PRIVATE LIFE

As Philippe Ariès and Georges Duby have shown in *Histoire de la vie privée: De l'Empire romain à l'an mil*,[15] privacy is a concept as old as the world and manifests itself in different forms in different cultures. However, the meaning we refer to today originates from an article that was published in 1890 by Samuel Warren and Louis Brandeis in the *Harvard Law Review*. It means "the right to be alone."[16] In other words, the right not to be disturbed by another person, without permission. It is also the right to be able to exercise control over the collection, use, and even storage of the data one produces. The web giants have often wanted to fight this right to privacy. Eric Schmidt, ex-CEO of Google, said this: "If you have something that you don't want anyone to know, maybe you shouldn't be doing it in the first place."[17] According to Schmidt, You have nothing to hide. No interest in having that right to privacy. You can share anything, even health information. Social media has picked up on this motto. Citizens and especially Politicians who share Eric Schmiidt's vision are now publishing a lot of information about themselves. There is nothing to hide.

Many recent examples contradict the view of the former Google CEO. For example, 45 minutes after posting a photo of his plane ticket on Instagram, the passport, and phone number of Tony Abbott, former prime minister of Australia, was found by a hacker.[18] The information we share online is valuable. It can generate precise knowledge about us that can help companies offer us personalized recommendations. A copy of your data (that you have nothing to hide) on Facebook contains 48 categories of information such as messages, posts, polls, events, payments, comments and reactions, stories, reels, groups, etc.

The company that holds this knowledge about us also holds power over us. Carissa Véliz calls for taking back control of our data because, according to her, we are a source of power.[19] But how can we exercise this control over our data?

There are several ways! First, when we share little or no data online. This goal is rather difficult to achieve because most of the basic services we use every day require the sharing of personal data.

Second, it is desirable to collect, store, and exploit the data we produce in a secure way. Moreover, any access to the data collected and stored must go through the person who produces it himself. Digital identity allows one to better operationalize this need to take control of one's data.

Third, when organizations that collect, store or use the data we generate inform us or ask for our consent to use our data. When it comes to consent, it must be revocable at any time and easily. Blockchain technology makes this revocation[20] and the traceability management of consent possible.

Fourth, it is to limit the share of our data to a trusted third party that can manage the rights to manage our data for our benefit. Public organizations, associations, or legal vehicles such as trusts may fit this scenario.

SENSITIVE DATA RAISES PRIVACY ISSUES

Sensitive personal data that is generated directly by individuals in areas, such as health, finance, and social media, must be preserved. It may share information that an individual may not want to know about him or her such as health information, financial profile, political ideology, etc. Similarly, the presence of IoT in our daily lives and their impact on health, finance and critical infrastructure requires the preservation of our privacy. The use of data generated by individuals and the IoT raises many issues from a privacy perspective. It is these issues that we want to raise by publishing this book which is divided into three sections.

The first section will address digital identity as a technological and legal vehicle for personal data protection. In the chapter "Blockchain for Digital Identity," the author proposes an overview of self-sovereign identity. He demonstrates its relevance for the transfer of ownership to the user as well as several use cases.

The second section considers privacy issues in the context of IoT and mobile payment systems. The chapter "Security and Corporate Violation to Privacy in The Internet Of Things Age" focuses on the threats coming from, on the one hand, the online networks provided by these sensors and, on the other hand, the tracking and profiling risks of the mobile computing environment. It analyzes the impact of IoT on individuals and existing data privacy regulations.

The authors of the chapter "Security, Privacy and Blockchain in Financial Technology" present the main existing implementations and solutions of mobile payment systems. They describe the ethical issues of personal data security and the role of blockchain in protecting privacy.

Finally, the last section discusses the challenges of dealing with sensitive data. The chapter "Where does the novel legal framework for AI in Canada stand against the emerging trend of online test proctoring?" focuses on the legal issues of socioeconomic discrimination and privacy of AI application in education. The authors assess the risks generated by AI tools for exam monitoring and the Canadian legal framework for data protection legislation and artificial intelligence (Bill C-27) in comparison with the European Commission's proposed AI act.

The next chapter focuses on double data spending, or the unauthorized secondary use of individuals' data, "Blockchain, AI and Data Protection in Healthcare: A Comparative Analysis of Two Blockchain Data Marketplaces in relation to Fair Data Processing and the 'Data Double Spending' Problem." The authors propose an articulation of fair data processing based on the Hyperledger Indy/Aries protocol and Ethereum.

The last chapter "Cyber influence stakes" focuses on the main techniques used by social media to influence the population. The others summarize studies related to the detection of trolls, bots and deepfakes to eradicate misinformation.

NOTES

1. Planetoscope, *Informations publiées dans le monde sur le net (en Gigaoctets)*, 2022. https://www.planetoscope.com/Internet-/1523-.html#:~:text=Quel%20volume%20d'informations%20produites%20chaque%20jour%20%3F&text=En%202018%2C%20on%20estime%20que,%2C5%20quintilions%20d'octets, accessed on September 11, 2022.
2. Lionel Sujay Vailshery, "Number of Internet of Things (IoT) Connected Devices Worldwide from 2019 to 2021, with Forecasts from 2022 to 2030", *Statista*, August 22, 2022. https://www.statista.com/statistics/1183457/iot-connected-devices-worldwide/#:~:text=The%20number%20of%20Internet%20of,around%205%20billion%20consumer%20devices, accessed on September 11, 2022.
3. Commission européenne, *Règlement sur les données: la Commission propose des mesures en faveur d'une économie des données équitable et innovante*, 23 février 2022, Bruxelles. https://ec.europa.eu/commission/presscorner/detail/fr/ip_22_1113, accessed on September 10, 2022.
4. J.-Y. Kim, C.-H. Chu and M.-S. Kang, "IoT-Based Unobtrusive Sensing for Sleep Quality Monitoring and Assessment," in *IEEE Sensors Journal*, vol. 21, no. 3, pp. 3799–3809, 1 February 1, 2021, doi: 10.1109/JSEN.2020.3022915.
5. https://en.wikipedia.org/wiki/IEEE_802.11
6. Risk prevention and control, *Référentiel APSAD R7 Détection automatique d'incendie Règle d'installation et de maintenance*, 2021. https://cybel.cnpp.com/livre-referentiel-apsad-r7-detection-automatique-d-incendie-2021,, accessed on September 11, 2022.
7. S. Bahizad, "Risks of Increase in the IoT Devices," 2020 7th IEEE International Conference on Cyber Security and Cloud Computing (CSCloud)/2020 6th IEEE International Conference on Edge Computing and Scalable Cloud (EdgeCom), 2020, pp. 178-181, doi: 10.1109/CSCloud-EdgeCom49738.2020.00038.société sont importants.
8. S. Bahizad, "Risks of Increase in the IoT Devices," 2020 7th IEEE International Conference on Cyber Security and Cloud Computing (CSCloud)/2020 6th IEEE International Conference on Edge Computing and Scalable Cloud (EdgeCom), 2020, pp. 178–181, doi: 10.1109/CSCloud-EdgeCom49738.2020.00038.
9. Alberto Huertas Celdrán, Pedro Miguel Sánchez, Fabio Sisi, Gérôme Bovet, Gregorio Martínez Pérez, Burkhard Stiller, Creation of a Dataset Modeling the Behavior of Malware Affecting the Confidentiality of Data Managed by IoT Devices, Robotics and AI for Cybersecurity and Critical Infrastructure in Smart Cities, 10.1007/978-3-030-96737-6_11 (193–225) (2022).
10. https://www.technologyreview.com/2021/08/30/1033941/afghanistan-biometric-databases-us-military-40-data-points/
11. https://theintercept.com/2021/08/17/afghanistan-taliban-military-biometrics/
12. D51, p. 10. https://eur-lex.europa.eu/legal-content/EN/TXT/PDF/?uri=CELEX:32016R0679&from=EN
13. https://www.vie-publique.fr/en-bref/283970-internet-des-objets-rapports-humains-machines
14. Government of Canada, *Research Security Policy Statement – Spring 2021*, 2021. https://www.canada.ca/en/innovation-science-economic-development/news/2021/03/research-security-policy-statement--spring-2021.html, accessed on September 14, 2022. Chaire Raoul-Dandurand, *Geopolitical Cyber Incidents in Canada*, 2022, p.7. https://dandurand.uqam.ca/wp-content/uploads/2022/08/2022-08-22-Rapport-OCM-ENG.pdfl, accessed on September 19, 2022.
15. Philippe Ariès et Georges Duby dans *Histoire de la vie privée : De l'Empire romain à l'an mil*, Paris, Seuil, 1999.
16. Samuel D. Warren; Louis D. Brandeis, "The Right to Privacy", *Harvard Law Review*, Vol. 4, No. 5. (Dec. 15, 1890), pp. 195. http://links.jstor.org/sici?sici=0017-811X%28189012 15%294%3A5%3C193%3ATRTP%3E2.0.CO%3B2-C, accessed on September 17, 2022

17. https://www.eff.org/fr/deeplinks/2009/12/google-ceo-eric-schmidt-dismisses-privacy
18. Derya Ozdemir, "Former Australian PM Hacked after Sharing Boarding Pass on Social Media" *Interesting Engineering, Sep 18, 2020.* https://interestingengineering.com/culture/former-australian-pm-hacked-after-sharing-boarding-pass-on-social-media, accessed on September 17, 2022.
19. Carissa Véliz, Privacy is Power: Why and How You Should Take Back Control of Your Data, p. 49.
20. Schallum Pierre, "Blockchain et santé. : cas d'utilisation pour l'autogestion de la maladie chronique", *Les techniques de l'ingénieur,* IN241, v1, 10 février 2021, p. 6.

Section I

Digital Identity Era

1 Demystifying the Digital Identity Challenges and the Blockchain Role

Manel Grichi
Department of Computer and Software Engineering, Polytechnique Montréal, Québec, Canada

Fehmi Jaafar
Department of Mathematics and Computer Science, Québec University at Chicoutimi, Québec, Canada

CONTENTS

1.1 Introduction 9
1.2 Digital Identity Challenges and Risks 11
  1.2.1 The Trust Challenge 11
  1.2.2 Identity Provider a Single Point of Failure 11
  1.2.3 Increased Phishing Risk 11
  1.2.4 Privacy Issues and Data Breaches 12
  1.2.5 Identity Fraud 12
  1.2.6 Lack of Reusability of Identities 13
1.3 Background 13
1.4 Literature Review 14
1.5 Current Experiences Through the World 17
1.6 Proposed Solution 18
  1.6.1 Architecture of the Proposed Solution 18
    1.6.1.1 Application Layer 18
    1.6.1.2 Blockchain Layer 20
    1.6.1.3 Database Layer 20
1.7 Conclusion 21
References 22

1.1 INTRODUCTION

Citizen identity information has become critical and significant information that plays a major role in maintaining relationships between citizens and the state [1]. Citizens are recognized by different attributes that give them meaningful and

DOI: 10.1201/9781003227656-3

unique identification such as name, national ID number, address, national health number, student number, passport number, driving license number, birth number, and social security number. Digital identity refers to the association of these attributes with citizens in a digital world that may be possible with Information Communication Technologies (ICT) [2].

These citizen identity information parameters help in managing access and the usage of a vast range of public and consumer services that include public government services, e-commerce, e-health facilities, e-voting, and toll tax. The citizens can have different identities simultaneously; they may have different documents as proof of identity such as passport, birth certificate, student card, or even health card. In order to take leverage of those identities, the citizens may need to visit those public departments of services in person and wait for their turn in the long queues. Now, the governments have extended their customer or public services and improved their public relations and this has been made possible with the help of digitization and ICT [3, 4]. The goal is that all these tasks can be performed remotely with the usage of internet-based technologies, a vast number of services, electronic payments, toll roads, or speed cameras, e-commerce, e-voting, and other forms of electronic services, based updates in case of emergencies or natural disasters, etc. All these services and information exchange between the public and the government are made possible by storing, managing, processing, and updating the information of the citizens, which is known as the citizen digital identity [3, 4]. Thus, the protection of the digital identity of the citizens is an extremely important and challenging task.

Given the significance of digital identity, many developed countries such as China, Estonia, India, Australia, the United States, and the United Kingdom are not only making the digital identity scheme as part of their e-government tasks and initiatives but also using the same identity for both public and private transactions [5]. These countries have been using citizen digital identity in a wide range of programs such as the Estonian e-residency program which is the first government-authenticated and operated international digital identity program for individuals who are neither Estonian citizens nor residents, or even physically present in Estonia, and the EU digital single market, etc. China is also trying to implement and use blockchain technology to implement the concept of digital identity. In 2016, China established the "Guangdong Province Big Data Comprehensive Experimental Area" in the Chancheng District, Foshan City of Guangdong Province [6]. Moreover, a number of public and private organizations have introduced different citizen digital identity management solution and the use of blockchain technology is among the most successful and most widely proposed solutions [7].

Systems based on the management of digital identity have provided new ways of doing business, however, there are also some implications associated with it in the form of identity fraud such as the creation of new false digital identities, fraudulent transactions, and money laundering, etc. [5, 8]. Nowadays, all the records and information of citizens are present in the computers even the social security, credit cards, medical records, and any other electronic transaction record. The misuse of the digital identity of any citizen should be considered theft of identity [5].

1.2 DIGITAL IDENTITY CHALLENGES AND RISKS

Digital identity presents several benefits in different application areas as discussed in Section 1. Despite the huge opportunities that it offers, there are still some challenges and risks that are associated with it such as scalability, security and privacy, compliance, and governance issues that still need to be explored and solved. Digital identity presents risks of spying, manipulation, and identity theft. Because of the amount of identity information stored and managed by organizations, security is an essential requirement for the service provider [9]. Current identity management systems have implemented methods, techniques, and frameworks to securely handling identity; however, there are still some vulnerabilities.

1.2.1 The Trust Challenge

Identity management systems require that the user and the relying party have trust in the identity provider because a lot of identity information is stored at identity providers and users can do nothing. Simply, trust them to preserve their privacy and properly secure their identity information. Sometimes, privacy-sensitive information can become public due to any human or technical error. Indeed, the possible large collection of data stored at an identity provider can also be used to commit identity fraud. If the information about a user that is stored at the identity provider becomes public due to any reason such as theft, hacking, or implementation flaws, then this data can be used to fake an identity when registering for a new service. To prevent this risk, it is necessary to put the user in control of information that is released from the identity provider, not only by policy but also technically enforced into the identity management system. It should not be possible for the identity provider to log in to a relying party claiming to be another user [10].

1.2.2 Identity Provider a Single Point of Failure

A single point of failure is a potential risk in centralized digital identity systems. In fact, the identity provider can be a single point of failure if it is specified as the only one system with specific responsibilities such as authentication management. A technical problem in the identity provider system can disrupt all the digital identity management systems. Thus, the distributed nature of blockchain technology can overcome the single point of failure, so the digital identity management system is less likely to experience downtime.

1.2.3 Increased Phishing Risk

Most current identity management systems only provide a way to authenticate the user but it is not possible for the user to authenticate the identity provider or the relying party. It is very important to prevent phishing attacks where attackers trick or manipulate users into revealing identity data and credentials. With the widespread use of identity management, phishing attacks based on getting identity providers' login credentials will most likely increase as well. Phishing attacks are much easier to occur when HTTP redirects are used [10]. It is as simple as creating an illegitimate but attractive website that redirects to a false copy of the identity provider to capture the user's credentials. A very common example of phishing attacks using fake identity provider websites is Yahoo! sign-in seal[1]. The presence of the seal enables the user to distinguish the real Yahoo! sign-in page from a false one. However, this solution only works if the user logs in using the same computer as the seal was created on, as Yahoo! identifies it by storing tags in multiple places on the computer. In order to prevent phishing attacks, it is very important that users can authenticate the relying party and the identity provider. Mutual authentication needs to be included in identity management systems in such a way that the user is not required to install special software or to use one and the same computer all the time. Concretely, the use of digital certificates can solve this problem as trusted certificates and enforce https protocol to mitigate these kinds of attacks. Moreover, the authentication of the identity provider and relying party by the user should be more user-friendly than checking their SSL certificate manually (SSL is a Secure Sockets Layer Protocol for establishing secure links between networked computers) [10].

1.2.4 Privacy Issues and Data Breaches

Sharing personally identifiable information is a great concern in managing privacy, protecting data, and complying with regulations. In identity management systems, sharing such information is often a key goal, which raises interesting privacy issues. It is possible, for example, for a service provider site to learn a user's globally unique digital identifier during the process, even if it's not necessary to know who the user is. "Pseudonyms" have been proposed for preserving privacy, especially when multiple web services cooperate to provide a combined service that necessitates user-attribute sharing [11]. Dunphy *et al.* [12] also reported that, recently the largest data breach of the 21st century occurred at Equifax with 143 million compromised identity data records. Recent studies showed that (1) nearly 60 million Americans were affected by identity theft in 2017, (2) more than 10 billion records have been breached since 2013, (3) over 6,500 incidents resulted in compromised data were publicly disclosed in 2018, (4) the average cost of identity fraud is estimated to $263 per person, (5) The yearly total cost of identity theft was estimated in 2016 to 16$billion, (6) the average cost for each stolen or lost record containing sensitive and confidential data is estimated to 148$. All those facts require developing novel identity solutions that strive toward a return to a satisfactory level of privacy [13].

1.2.5 Identity Fraud

Financial identity fraud is the most common type of identity fraud[2]. However, Identity fraud can happen in many forms such as medical identity or the creation of a new digital identity. The goal of identity fraud is to create fraudulent identities using fake or real information, or a combination of the two. Then, this identity can be used in online or offline criminal activities.

$16 billion was stolen from 15.4 million US consumers in 2016, compared with $15.3 billion and 13.1 million victims a year earlier [14].

1.2.6 Lack of Reusability of Identities

Reuse of digital identities has great potential to reduce the data replication and environmental footprint of the data. However, the way digital identities are specified and used rarely considers their reuse in various applications and information systems. This fact will create costs for organizations, lack of interoperability, and usability challenges for users. Thus, it is necessary to identify challenges and success factors for reuse of digital identities in pilot projects, to identify the technical and regulation challenges to optimize the reusability of identities, and to select the best practices that foster the reuse of digital identities.

Given the challenges and risks in the digital identity systems, several researches proposed the use of blockchain technology as their solution in different capacities. Since blockchain is a decentralized system, it does not need a third-party trusted authority. Rather, it adopts the decentralized consensus mechanism in order to guarantee the reliability and consistency of data and transactions.

1.3 BACKGROUND

Blockchain is a digital ledger of economic transactions continually updated by several users publicly with a focus on maintaining the integrity of transaction data, It is a chain of continuous records in blocks and it originated with the introduction of the bitcoin cryptocurrency [15]. Blockchain technology has transformed business technology and the way of doing business transactions with security and integrity. Business organizations have been gaining a competitive advantage over their competitors by leveraging this technology. A typical blockchain database consists of two types of records, that is, transactions and blocks where blocks are responsible for holding batches of transactions. The blockchain has four elements that are replicated: the ledger, cryptography, consensus, and business logic [16].

Blockchain was initially introduced with the emergence of bitcoin cryptocurrency for providing a secure and reliable source of digital transactions. In bitcoin, blockchain provides a decentralized technique for transferring money for the users. It also provides an environment for digital contracts and peer-to-peer data sharing in a cloud service [17]. However, bitcoin cryptocurrency is not the only technology that has been using blockchain, it has now been widely used in various domains such as finance, business, health, smart cities, and maintaining digital identity in government organizations [18, 19].

Blockchain is a decentralized system, and it does not need a third-party trusted authority. Rather, it adopts the decentralized consensus mechanism in order to guarantee the reliability and consistency of data and transactions, the current blockchain mechanism has four major components: Pow (proof of work), Pos (proof of stake), PBFT (practical byzantine fault tolerance), and Dpos (delegated proof of stake). For example, the two most common systems based on blockchain, that is, Bitcoin and Ethereum are using the Pow mechanism [20].

Blockchain provides security and privacy in way that it provides a way of digital identity so that consumers and users feel more comfortable in sharing their data. This self-sovereign identity is based on two principles *i.e.,* consent and control. Consent is the permission among individuals and organizations defining which private information could be used by, while, the control feature ensures the complete ownership of personal data to their owners. This self-sovereign identity model gives privacy control to the consumer as it reduces the probability of identity breaches and fraudulent activities in businesses [19].

Having said that the world is digitally connected today and every transaction of data can be performed within seconds across the globe, however, there still exists the question of security of data and transactions. Blockchain technology for managing digital identity enables citizens to build a more connected society with a secure identity. Another main advantage of blockchain is that the public ledger cannot be modified or deleted after the data has been approved by all nodes. Due to these primary features of security, data integrity, and anonymity without the need for a third-party entity/organization, blockchain has its significance in digital identity [20]. The blockchain works assuming that honest and reliable nodes control the whole network and if the attacker nodes get more computational power than the honest nodes, then the network could become vulnerable to attacks [17, 20].

Since, blockchain technology does not depend on centralized control authority, it can efficiently manage both consent and control of personal information. It contains smart contracts and associated rules for securing the personal information of consumers. They define who can collect identity-related data, who has access to it, and to what level that access is granted [19]. For example, blockchain can verify identity without revealing details behind that identity. In other words, required data can be widely shared in a transparent manner and protected at the same time.

1.4 LITERATURE REVIEW

With the rapid transformation of business processes and their emerging requirements to meet the challenging markets, businesses have started adopting ICT-based solutions where the concept of digital identity is no exception. The use of digital citizen identity has not only made business processes efficient and effective but also it has provided a way of security and privacy of personal data and information. Several researches have been conducted on digital citizen identity where they discussed its usage, its benefits in different application areas, and the prospective that it has given.

Blockchain has been an emerging technology that is going to lead the modern world with its revolutionary technology due to the features it provides such as security and transparency. Many researchers have proposed the use of blockchain technology in digital identity to make it more secure and for avoiding the violation of privacy. AlMamun *et al.* [21] proposed a system on blockchain-based citizen digital identity using bio-information. They implemented their proposed digital identity system using Ethereum smart contract. Smart contracts are a type of Ethereum account. This means they have a balance and they can send transactions over the network.

However, they are not controlled by a user, instead they are deployed to the network and run as programmed[3]. The results of AlMamun *et al.* show that an attacker cannot access the personal information of a citizen and any unauthorized access attempt is denied instantly, thereby ensuring the privacy of private citizen data. Same as AlMamun, Sin and Naing [22] also proposed a digital identity management system to decrease duplicate identity documents. Htet *et al.* [23] proposed the use of blockchain technology for digital identity management systems as a solution for passport digital identity in Myanmar. Their proposed system enabled to identify and control illegal duplication of passports. Since, a valid passport can be issued only once to a person (except passport loss, renewal, and expired) and it is impossible for a person to apply for more than one passport type in the decentralized system. Dong *et al.* [24] proposed a solution for digital identity management and securing the information of banking customers. They proposed BBM which is a blockchain-based self-sovereign identity system model for open banking. Self-sovereign identity (SSI) is a model for managing digital identities in which an individual or business has sole ownership over the ability to control their accounts and personal data. With an SSI approach, the users have complete control over how their personal information is kept and used[4]. Their proposed model provides a secured way for open banking customers who hesitate in sharing their personal information with third-party service providers. Their model makes it possible to have a reliable and secure connection between open banking customers and third-party service providers by allowing the customers to manage and control their own identity and data. Li *et al.* [25] also used blockchain technology in the context of court trials and evidence management. They proposed a blockchain-based event management scheme LE-chain that supervises the complete flow of evidence and all of the court data such as votes and trial results, etc. They use short random signatures that anonymously authenticate the identities of witnesses to protect their privacy. Noack and Kubicek [26] discussed the introduction of digital identity for online authentication in Germany. Germany started the process of electronic-ID in the late 1990's incorporating e-governments and e-commerce where they implemented a legislation on e-signatures that was supposed to be used for the online authentication of citizens. This was then replaced by e-ID card in 2009 that was considered as a radical innovation that provided a two-sided authentication to the citizens.

Healthcare has also been a very important area where the identity of the citizens is crucial. Many researchers discussed the importance of digital identity in health care and how it could be used to secure the patients' data and to improve their user experience. Christine [27] proposed the use of blockchain-based digital platform to serve as a secure patient data repository while Bhuiyan *et al.* [28] proposed a blockchain-based solution for managing and sharing the patients' healthcare data across different medical centers/hospitals in a secure way. Several other researchers have also proposed the effective and secure management and sharing of healthcare data among different stakeholders.

Many countries have been adopting the digital identity of citizens when it comes to the healthcare sector. For example, in Canada, that is, the policymakers, public leaders, corporate leaders, and entrepreneurs are making significant efforts to help the country be the leader of the next era of the Internet as a platform that helps transform human affairs for the benefit of the citizens [7]. McEachern and Cholewa [29] reported that the Government of Alberta is making efforts in allowing its citizens to access their health information online and Alberta is the only province in Canada that provides provincial electronic health records to its citizens and practitioners. The integrity, availability to the right user, and accuracy of data are the most important parameters. Service Alberta developed the My Alberta Digital ID program for providing a digital identity to the citizens. The Government of Alberta aims to provide its citizens with reliable and secure healthcare information maintaining their digital identity. Wolfond [7] discussed the need to implement a digital identity using blockchain technology in Canada. He argues that with an inclusive, comprehensive, and secure approach to identification, Canadian healthcare could be significantly transformed. This could enable streamlining patient administration, engaging consumers in self-care and management at home, and supporting those who manage the wellness of their families. Using digital identity, patients and providers could securely perform their tasks such as the identification during appointment bookings, access records, and authorize a "circle of care" to share their patient history across multiple providers and family members.

Financial sector is one of those business areas where identity of individuals or citizens has a big importance. Millions of financial transactions are performed daily in every sector such as banking, e-commerce, trading, real-estate, transportation, health, education, manufacturing industry, and automobile industry. In all these sectors, the identity of citizens plays a critical role and it could cause huge disasters if not managed carefully. Researchers have also proposed several approaches for using digital identity in this area. Wolfond [7] proposed an approach that leverages the benefits of both blockchain and digital ecosystem (i.e., a group of interconnected information technology resources that can function as a unit) within the Canadian context. They further argue the usage and benefits of using digital identity in the economic and financial sectors that they are trusted across business organizations and must be used to allow users to prove their identity in a secure and privacy-enhancing way. This digital citizen identity would protect users against the increasing rate of cyber frauds and cybercrimes by increasing the trust and safety of citizens. Al-Khouri [30] discusses the importance of digital identities in the economic growth of business and government organizations. The author presents an overview of the identity management infrastructure development initiatives in Gulf Cooperation Council countries. He also examines their potential to revolutionize and transform existing economic models. He argues that the smart identity management system in these countries may transform their business transactions by creating a trustworthy environment and providing them with a secure digital identity that would finally give rise to the digital economy. Angelakopoulos and Mihiotis [31] discussed in their paper the challenges and opportunities of e-banking in Greece. Their results show that in order to expand to e-banking services, gain a competitive advantage, and meet the technological requirements and challenges, this adoption of digital identity in the banking sector should provide a trustworthy and secure way of doing business and transactions without any hesitation or fear of losing personal data.

Another important application area of digital identity is the manufacturing industry, which is considered an important area of research in industrial automation. Using the concept of Industry 4.0, many industrial sectors are making efforts to improve and advance their systems to achieve higher productivity, effectiveness, reliability, improved quality, and flexibility in their production. Achieving all these features of smart manufacturing causes several challenges such as security, trust, reliability, traceability, and agreement automation within the manufacturing value chain process, hence, Mohamed and Al-Jaroodi [32] proposed a blockchain-based digital identity system not only for people but also for different entities in the Industry 4.0.

1.5 CURRENT EXPERIENCES THROUGH THE WORLD

Digital citizen is the concept of giving identity to the citizens so that they can get their daily tasks done with much more ease, security, and effectiveness from anywhere. These tasks could be buying a bus pass, checking the status of a building under construction, paying taxes, receiving funds, and getting registered to any consumer service, getting a passport or a driving license. All these activities could be performed from one place or using a single identity which we call the digital identity. According to the report by Lynch [33], more than one billion people in developing countries around the world have no proof of their identity. Thus, the unique digital citizen identity would bring interesting changes in every process of daily life and provides identity proof [34]. A major example is the Estonian advanced model of digital identity that enables every government sector to identify its citizens with their unique digital identity to facilitate them with basic needs. They have the facility of online voting, access to their healthcare records online, and can access all the government processes online such as marriage, divorce, real-estate, and any other registration or record checking, etc. Other than the developed countries such as Canada, the United States, the United Kingdom, and other European countries, a number of developing countries including Peru, Chile, Pakistan, India, Thailand, Indonesia, and the Philippines, have also been trying to take the leverage of digital citizen identity [34]. Bitnation a governance platform based on blockchain technology, tries to establish the concept of world-citizenship through digital identity registration on blockchain [35]. Any individual in the world can become the citizen of Bitnation by agreeing to the constitution. The Bitnation Refugee Emergency Response (BRER) program provides an identification system called Blockchain Emergency ID (BE-ID). There are a number of programs, organizations, and software that provide the facility of having a digital identity including ConsenSys, ID2020, Hyperledger, Australia Post Office, PIMN (Platform Identity Management Netherlands), The US Homeland Security, ShoCard, and Uport [36].

Other than these applications of digital identity, Canada has emerged as a country that is rapidly adapting to life in the digital age. The citizens have been leveraging these technologies in every area of life such as shopping, banking, and accessing government services. A report by the Digital Identification and Authentication Council of Canada (DIACC) [37] highlighted the need for having a new digital identity system that would allow citizens to perform all their tasks in person. The digital identity would be used to control documents by both the federal government and the provincial government in the public and private sectors. The report also says that the new digital identity system must be robust, secure, scalable, and provide no additional risk to personal information and privacy. Sunil Abraham [38] discusses Canada's model for a digital identity that incorporates multi-stakeholder coordination, network paradigm, following of standards not technology, clarity on intellectual property, embrace of the latest technology, interoperability and compatibility, and soft infrastructure of digital identity. The author argues that other countries should also look into the Canadian model and adopt it for implementing a digital identity management program. They need to build trust in a comprehensive way that the people themselves become champions of the digital identity ecosystem.

1.6 PROPOSED SOLUTION

1.6.1 Architecture of the Proposed Solution

We propose in Figure 1.1 the architecture of a secure digital identity application. We define three main necessary layers. Our goal is (1) to ensure citizen data privacy and get it protected with a secure blockchain network, (2) to give to the citizen the control of his identity and his data, and (3) to allow a quick recovery of his identity.

As a concrete example, we present in Figure 1.2 a use case of the renewal of the health insurance card 2.

Figure 1.2 shows three main actors, that is, the citizen, the trusted authorities, and two main government services, that is, Health insurance service and financial service. The financial service is used by the blockchain module to verify and validate the digital identity of the citizen who requests the renewal of his health card without accessing to his personal financial information. This verification requests the approbation of the citizen himself who has to authorize the verification of his identity through another service.

1.6.1.1 Application Layer

The application layer is the first module in this architecture, it contains the different API of the government public services where any citizen could access it online and request specific services *i.e.*, renewal of health insurance card, renewal of passport,

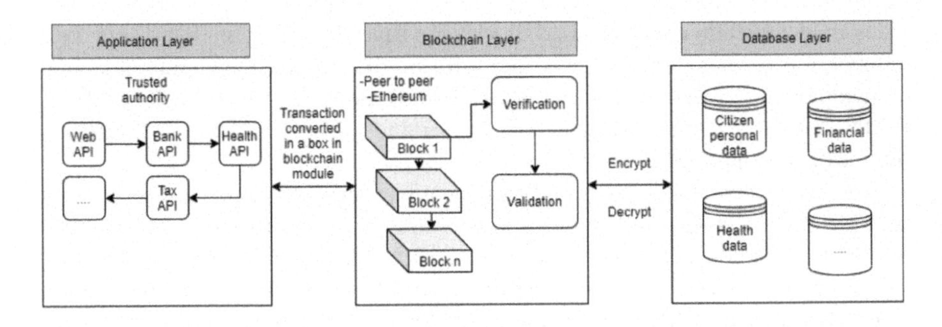

FIGURE 1.1 Architecture of the proposed solution.

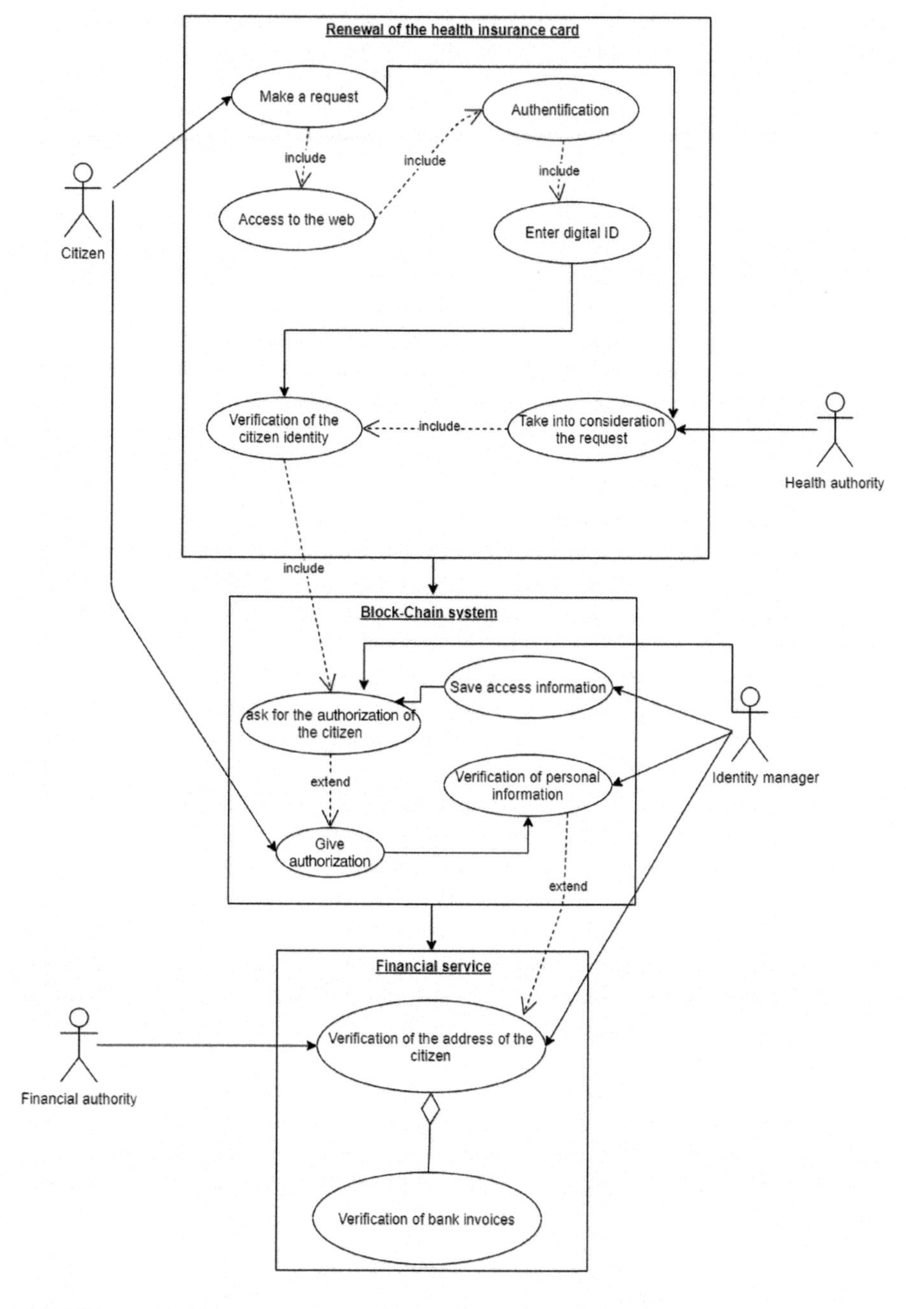

FIGURE 1.2 UML use case – renewal of the health insurance card.

request a credit card, etc. It is considered as an interactive layer with the citizen as it is the closest step to the citizen, that is, the user can easily manipulate his request via the web.

Thus, the application layer specifies the communications protocols and interface methods used in digital services and ensures a stable communication across the architecture.

Concretely, the application layer handles the following functions:

- It ensures that the device used by a citizen to receive digital services is identified and ready to start the authentication process.
- It enables authentication between the device and the service provider.
- It establishes rules and practices in case of connectivity problems such as error recovery procedures, data integrity verification, etc.

1.6.1.2 Blockchain Layer

The Blockchain layer is the module responsible for the citizen's information security and for the verification of his identity, that is, the user who interacted with the application layer to request public services. This layer is the second module in our architecture placed behind the application layer and which is hidden from the citizen. Once the citizen interacts with the application layer to ask for a service, his request will be converted into a block that will be incarnated into the blockchain module and then shared between all his nodes. This layer ensures the verification and then the validation of the citizen's digital identity once he is authenticated to public government service.

Several researchers have proposed approaches that use blockchain technology to address security and privacy issues in digital identities. In fact, blockchain is an immutable distributed ledger, able to record transactions that are occurring within a network. All participants maintain a copy of the ledger and reach a consensus on the states of transactions submitted to the blockchain network [39]. The benefits of blockchain-based techniques in digital identities management systems are various, and we can cite the fact of removing the singular point of failures and preventing third parties from controlling data. Thus, participants can verify the data integrity and identity sender.

We propose to use public a blockchain as one of the characteristics of a public blockchain is being an open-access platform [40]. In other words, any node can read and record transactions. Such aspects prevent public blockchain from being hacked because is expensive to add a new block and tamper the blockchain content [41].

1.6.1.3 Database Layer

The database layer is the third module in our proposed architecture. It is the layer responsible for storing citizens' personal data to ensure its privacy. It contains different databases to provide various digital services *i.e.,* financial database, health database, etc. Each one contains specific information related to the service itself. These different databases could communicate together or share citizen personal information with the approbation and authorization of the citizen himself. The management of the database layer includes the use of a set of security services dedicated to preventing data breaches such as data breach prevention or data loss prevention (DLP). Such services focus on the integrity and nonimpregnability of these databases themselves. Concretely, in this layer, the focus must be on securing the digital identity using the zero trust principle. This principle implies the application of a set of security rules in accordance with policy and permissions. As an example of rules applied in the data base layer of digital identity management systems, we can enumerate the following points:

- Who has access to what data related to digital identities?
- Which application is accessing digital identity data?
- When are digital identities accessed?
- How are digital identities data moving to service providers?
- Why is this user accessing to a digital identity data?

The goal behind the application of these rules in digital identity management system is to protect access to personal data and ensure that only authorized entities can have a view of such data.

1.7 CONCLUSION

The main concern of the government is to provide a secure software system to their citizens in order to ensure their personal information is private and avoid the risk of identity theft when they access the public government services provided online. In this paper, we highlighted the digital citizen identity concept that ensures to provide daily services accessible with much more ease, security, and effectiveness from anywhere. The goal is to provide a state of the art of the benefits, challenges, and the available solutions of this concept. Therefore, we performed a literature study where we studied the literature, and then we highlighted the experience of using the digital identity in different countries throughout the world. Based on that, we proposed an architecture of a solution that will help to overcome the challenges identified in our literature study. This architecture contains three layers *i.e.*, applications to give the citizen the control of his identity and data, blockchain to ensure citizen data privacy and get it protected, and database to enable the secure access to the citizens' data with more ease and to allow an identity quick recovery.

NOTES

1. https://www.webdevelopersnotes.com/what-is-yahoo-sign-in-seal
2. https://www.equifax.com/personal/education/identity-theft/types-of-identity-theft/
3. https://ethereum.org/en/developers/docs/smart-contracts
4. https://searchsecurity.techtarget.com/definition/self-sovereign-identity

REFERENCES

[1] M. Lips, "Rethinking citizen – government relationships in the age of digital identity: Insights from research," *Information Polity*, vol. 15, pp. 273–289, 12 2010.

[2] L. Camp, "Digital identity," *Technology and Society Magazine, IEEE*, vol. 23, pp. 34–41, 02 2004.

[3] M. Lips, J. Taylor, and J. Organ, "Identity management as public innovation: Looking beyond id cards and authentication systems," *Journal of Marketing – J MARKETING*, vol. 12, pp. 204–216, 01 2006.

[4] ——, "Identity management in e-government service provision: Towards new modes of government and citizenship," pp. 151–168, 01 2010.

[5] C. Sullivan, "Digital identity – from emergent legal concept to new reality," *Computer Law Security Review*, vol. 34, no. 4, pp. 723–731, 2018. [Online]. Available: http://www.sciencedirect.com/science/article/pii/S0267364918302024

[6] H. Hou, "The application of blockchain technology in e-government in China," in *2017 26th International Conference on Computer Communication and Networks (ICCCN)*, IEEE, Vancouver, Canada, 2017, pp. 1–4. Available: https://doi.org/10.1109/ICCCN.2017.8038519.

[7] G. Wolfond, "A blockchain ecosystem for digital identity: Improving service delivery in Canada's public and private sectors," *Technology Innovation Management Review*, vol. 7, pp. 35–40, 10 2017.

[8] K. M. Orzech, W. Moncur, A. Durrant, and D. Trujillo-Pisanty, "Opportunities and challenges of the digital lifespan: Views of service providers and citizens in the UK," *Information, Communication & Society*, vol. 21, no. 1, pp. 14–29, 2018.

[9] J. Torres, M. Nogueira, and G. Pujolle, "A survey on identity management for the future network," *IEEE Communications Surveys & Tutorials*, vol. 15, no. 2, pp. 787–802, 2012.

[10] G. Alpár, J.-H. Hoepman, and J. Siljee, "The identity crisis. security, privacy and usability issues in identity management," *arXiv preprint arXiv:1101.0427*, 2011.

[11] E. Maler and D. Reed, "The Venn of identity: Options and issues in federated identity management," *IEEE Security & Privacy*, vol. 6, no. 2, pp. 16–23, 2008.

[12] P. Dunphy, L. Garratt, and F. Petitcolas, "Decentralizing digital identity: Open challenges for distributed ledgers," in *2018 IEEE European Symposium on Security and Privacy Workshops (EuroS&PW)*. IEEE, London, UK, 2018, pp. 75–78. Available: https://doi.org/10.1109/EuroSPW.2018.00016.

[13] O. Dib and B. Rababah, "Decentralized identity systems: Architecture, challenges, solutions and future directions," *Annals of Emerging Technologies in Computing (AETiC)*, vol. 4, no. 5, pp. 19–40, 2020.

[14] A. Pascual, K. Marchini, and S. Miller, "Identity fraud: Securing the connected life," *Retrieved on January*, vol. 20, p. 2018, 2017.

[15] L. Carlozo, "What is blockchain?" *Journal of Accountancy*, vol. 224, no. 1, p. 29, 2017.

[16] P. Tasatanattakool and C. Techapanupreeda, "Blockchain: Challenges and applications," in *2018 International Conference on Information Networking (ICOIN)*. IEEE, Chiang Mai, Thailand, 2018, pp. 473–475. Available: https://doi.org/10.1109/ICOIN.2018.8343163

[17] J. Yli-Huumo, D. Ko, S. Choi, S. Park, and K. Smolander, "Where is current research on blockchain technology?—A systematic review," *PLoS One*, vol. 11, no. 10, p. e0163477, 2016.

[18] R. Rivera, J. G. Robledo, V. M. Larios, and J. M. Avalos, "How digital identity on blockchain can contribute in a smart city environment," in *2017 International smart cities conference (ISC2)*. IEEE, Wuxi, China, 2017, pp. 1–4. Available: https://doi.org/10.1109/ISC2.2017.8090839

[19] IBM Institute for Business Value, "Trust me: Digital identity on blockchain," 2017. https://www.ibm.com/downloads/cas/KAKYAKWG. Last visited 2023-05-20.
[20] X. Li, P. Jiang, T. Chen, X. Luo, and Q. Wen, "A survey on the security of blockchain systems," *Future Generation Computer Systems*, vol. 107, pp. 841–853, 2020.
[21] M. A. AlMamun, S. M. Alam, M. S. Hossain, and M. Samiruzzaman, *A Novel Approach to Blockchain-Based Digital Identity System*, in *Advances in Information and Communication: Proceedings of the 2020 Future of Information and Communication Conference (FICC), Volume 1.* Springer International Publishing, Switzerland, 02 2020, pp. 93–112.
[22] E. S. Sin and T. T. Naing, "Digital identity management system using blockchain technology," in *International Conference on Innovative Computing and Communications.* Springer, Germany, 2020, pp. 895–906.
[23] M. Htet, P. T. Yee and J. R. Rajasekera, "Blockchain based Digital Identity Management System: A Case Study of Myanmar," *2020 International Conference on Advanced Information Technologies (ICAIT)*, Yangon, Myanmar, 2020, pp. 42-47, doi: 10.1109/ICAIT51105.2020.9261785.
[24] C. Dong, Z. Wang, S. Chen, and Y. Xiang, *BBM: A Blockchain-Based Model for Open Banking via Self-sovereign Identity*, in *Blockchain–ICBC 2020: Third International Conference, Held as Part of the Services Conference Federation, SCF 2020*, Proceedings 3. Springer International Publishing, Honolulu, HI, USA, 09 2020, pp. 61–75.
[25] M. Li, C. Lal, M. Conti, and D. Hu, "LEchain: A blockchain-based lawful evidence management scheme for digital forensics," *Future Generation Computer Systems*, vol. 115, pp. 406–420, 02 2021.
[26] T. Noack and H. Kubicek, "The introduction of online authentication as part of the new electronic national identity card in Germany," *Identity in the Information Society*, vol. 3, no. 1, pp. 87–110, 2010.
[27] O. Christine, "Improving the user experience in healthcare through service design: Developing a digital identity for patients," Electronic Thesis or Dissertation, MDes - Master of Design, 2020-08-11T12:44:21Z, p. 175, 2020.
[28] M. Z. A. Bhuiyan, A. Zaman, T. Wang, G. Wang, H. Tao, and M. M. Hassan, "Blockchain and big data to transform the healthcare," ser. ICDPA 2018. New York, NY, USA: Association for Computing Machinery, 2018, pp. 62–68. [Online]. Available: https://doi.org/10.1145/3224207.3224220
[29] A. McEachern and D. Cholewa, "Digital health services and digital identity in Alberta," *Studies in Health Technology and Informatics*, vol. 234, pp. 222–227, 01 2017.
[30] A. M. Al-Khouri, "Digital identity: Transforming GCC economies," *Innovation*, vol. 16, no. 2, pp. 184–194, 2014.
[31] G. Angelakopoulos and A. Mihiotis, "E-banking: challenges and opportunities in the Greek banking sector," *Electronic Commerce Research*, vol. 11, no. 3, pp. 297–319, 2011.
[32] N. Mohamed and J. Al-Jaroodi, "Applying Blockchain in Industry 4.0 Applications," *2019 IEEE 9th Annual Computing and Communication Workshop and Conference (CCWC)*, Las Vegas, NV, USA, 2019, pp. 0852-0858, doi: 10.1109/CCWC.2019.8666558.
[33] S. Lynch, "Mission billion challenge," *Accessed: 2020-12-22*, 2020.
[34] D. Insights, "The digital citizen – improving end-to-end public service delivery via a unique digital identity," *Accessed: 2020-12-22*, 2020.
[35] B. Magazine, "Estonian government partners with bitnation to offer blockchain notarization services to e-residents," *Accessed: 2020-12-24*, 2020.
[36] O. Jacobovitz, "Blockchain for identity management," *The Lynne and William Frankel Center for Computer Science Department of Computer Science. Ben-Gurion University, Beer Sheva*, 2016.

[37] DIACC, "A report by the Digital Identification and Authentication Council of Canada," *Accessed: 2020-12-24,* 2015.

[38] S. Abraham, "Building trust: Lessons from Canada's approach to digital identity," 2020.

[39] S. K. Lo, Y. Liu, S. Y. Chia, X. Xu, Q. Lu, L. Zhu, and H. Ning, "Analysis of blockchain solutions for IoT: A systematic literature review," *IEEE Access*, vol. 7, pp. 58822–58835, 2019.

[40] X. Wang, X. Zha, W. Ni, R. P. Liu, Y. J. Guo, X. Niu, and K. Zheng, "Survey on blockchain for internet of things," *Computer Communications*, vol. 136, pp. 10–29, 2019.

[41] M. S. Ali, M. Vecchio, M. Pincheira, K. Dolui, F. Antonelli, and M. H. Rehmani, "Applications of blockchains in the internet of things: A comprehensive survey," *IEEE Communications Surveys Tutorials*, vol. 21, no. 2, pp. 1676–1717, 2019.

2 Blockchain for Digital Identity

Tim Weingärtner
School for Computer Science and Information Technology, Lucerne University of Applied Sciences, Rotkreuz, Switzerland

CONTENTS

2.1 Introduction 25
2.2 Identity Fundamentals 26
2.3 Use Cases 28
2.3.1 Cardossier – Live Cycle of an Automobile 28
2.3.2 Identity of Things 28
2.3.3 Findy 29
2.3.4 British Columbia – Digital Government 29
2.4 Identity Technology 29
2.4.1 Decentralized Identifier (DID) 30
2.4.2 DID Document 30
2.4.3 DIDComm 31
2.4.4 Verifiable Credentials (VCs) 31
2.4.5 Wallet 32
2.4.6 Zero-knowledge Proof 32
2.5 Governance 32
2.6 Critical Aspects of Self-Sovereign Identity 33
2.7 Conclusion 34
References 34

2.1 INTRODUCTION

Despite the advances in web technology over the previous few decades, there is currently no simple way to authenticate one's identity online. Unfortunately, when the Internet was created, the user's identity was lost. This did not have a significant influence at first. However, with the emergence of e-commerce, the flaw was immediately exposed. The phrase "On the internet, nobody knows you are a dog" does not originate from nowhere (Sovrin Foundation, 2018). Accounts and passwords were employed to fix the absence of identification, but this produced other issues. Meanwhile, private identity providers exploit the data for advertising and data research. As a result, in many nations, the development of an electronic identity that is not defined by individual players is required.

DOI: 10.1201/9781003227656-4

When it comes to identification, it is vital to remember that it is not simply about the information on a card; each individual has a variety of identities. This might be a university degree, a medical record, or membership in a club. Users' statements about their identities are almost impossible to prove on the web. However, without reliable proof, it is difficult to trust statements about a user's identity.

In the physical world, this is simply solved by reaching into one's wallet and showing an official document – for example, an identification document issued by a government authority, such as a passport, driver's license, or identity card. When boarding an airplane, renting a car, or borrowing a book from the library, identity is proven by a verification document issued by the representing authority. A corresponding digital equivalent hardly exists yet.

At the same time, digital and physical life are merging more and more. Our daily life gets tied to apps, services, and devices. The COVID pandemic in particular has significantly accelerated this digital transformation process. While such a transformation makes it possible to communicate with companies and other users to an extent that was previously unimaginable, it also makes it possible to communicate with other users on a more personal level. However, centralized solutions from private identity providers expose users to the risk that their data will not be carefully held by operators and could fall into the wrong hands in the event of data leaks.

Not only individuals need a verifiable and trustworthy identity. The rapidly growing number of devices (Internet of Things (IoT) devices) also require identification work and identity. The more security-critical areas are affected, the more important verifiable ownership, access authorizations, and location of such devices become.

Self-sovereign identity (SSI) is the concept of self-managing one's own identity characteristics. It not only involves names, address, date of birth, or nationality but also includes information about education, interests, families, sports activities, or club memberships. Much of this information is not verified by the state. For example, the university certificate is confirmed by the issuing university. With SSI, each person is in possession of their data. Confidence in this accuracy of the data is established through proofs provided by the respective issuers of the data. For example, the university is the issuer of the certificate and the academic title associated with it. These proofs are also called verifiable credentials. When the data is presented, the verifying body, entitled verifier, can ensure the authenticity of the data via the verifiable credentials. The trust triangle is established between *issuer*, *holder*, and *verifier*.

This chapter discusses SSI and its connection to blockchain technology. Following a review of identity fundamentals, many use cases explain how SSI may be applied. The technological foundations and standards are then examined in further depth. Furthermore, an important aspect, the governance, is highlighted. Finally, critical matters are addressed.

2.2 IDENTITY FUNDAMENTALS

Webster[1] defines identity as "the distinguishing character or personality of an individual." It comprises the totality of peculiarities that characterize an entity, object, or thing. This makes it distinguishable from others as an individual or unique instance. Identity attributes might be fingerprints, face, iris structure, voice, DNA, smell, speech, location as well as possession or access to physical objects like identity card, mobile device, notes, etc. Nonphysical attributes which define our identity mainly depend on our brain like knowledge, abilities, memories, experiences, relationships, feelings, wishes, behavior, or secrets. For an identity check, we compare those attributes with previously stored data.

Not only does SSI place the user at the center of the identification process but also requires that the user be in control of their own identity, including its data and how it is used. Based on these insights, Christopher Allen defines SSI in his article as follows:

> Sovereign identity is the next step after user-centric identity, and that means it starts at the same place: the user must be at the center of identity management. This requires not only interoperability of a user's identity across multiple sites, but also true user control over that digital identity, creating user autonomy. To achieve this, a self-sovereign identity must be transportable; it cannot be tied to one location or territory."
>
> *Allen (2016)*

Allen presents the ten principles of SSI (Allen, 2016):

(1) Existence. Users must have an independent existence.
(2) Control. Users must control their identities.
(3) Access. Users must have access to their own data.
(4) Transparency. Systems and algorithms must be transparent.
(5) Persistence. Identities must be long-lived.
(6) Portability. Information and services about identity must be transportable.
(7) Interoperability. Identities should be as widely usable as possible.
(8) Consent. Users must agree to the use of their identity.
(9) Minimalization. Disclosure of claims must be minimized.
(10) Protection. The rights of users must be protected.

SSI of people also in combination with blockchain technology is discussed in several research papers like in Mühle et al. (2018), van Bokkem et al. (2019), Stokkink and Pouwelse (2018), or Liu et al. (2017). In addition, IT enterprises step toward this direction like Microsoft Corporation (2018). In essence, SSI is about giving every user the power of disposal and decision-making over their own digital identity and with whom this data is shared – in other words, every person is their own sovereign over their digital identity. To enable this self-governance, it is necessary for the digital identity to be portable and decentralized and not dependent on any institution other than the user or owner of the identity. At the same time, however, it must be ensured that the person behind the digital identity is really the one they claim to be. The question of trust in such a claim is the central point which a functioning digital identity and the underlying system must fulfill.

This is where blockchain technology comes in. It plays a central role in the implementation of SSI. Blockchain acts as a trust anchor, that is, as an immutable, verifiable register in which the issuers' identity can be verified. It must be noted that SSI does not require a blockchain as trust anchor and works with a traditional registry. Nevertheless, a blockchain is the ideal complement since its immutable and decentralized character supports the philosophy behind SSI.

All the above apply with variations to IoT devices as well. As described in Weingaertner and Camenzind (2021), the concepts of SSI can be applied to IoT devices, which allows the traceability of the origin of these devices.

2.3 USE CASES

Following this overview, the following examples demonstrate the necessity of identification in a variety of digital scenarios. They demonstrate a wide range of potential applications for both blockchain and SSI. Many of these instances are on the cusp of becoming viable options. On the other hand, they are still in beta form and are closely linked to research.

2.3.1 Cardossier – Live Cycle of an Automobile

In 2017, the research project Cardossier[2] started in Switzerland. The aim of this project was to manage the life cycle of cars on a blockchain infrastructure. In 2019, the project members founded an association to continue the work of the initial research project. With the commissioning of the first 11 million records in 2020, a further important milestone was reached. Cardossier builds an infrastructure that allows all participants in the automotive ecosystem in Switzerland to implement their individual use cases and business cases on top. Examples are the car import process, car registration, used car dealership, or service documentation.

Identities of both – the participants and the cars – are a major aspect of this whole ecosystem. Due to the lack of an official electronic identity, Cardossier association implemented an SSI-based solution[3]. This first pilot shows the potential of the combination of existing government services like certification of residence and new, innovative solutions. It also shows some of the limitations SSI is facing today. First and foremost, the user experience and usability. Many wallets (see below) are still in a rudimentary stage and designed for savvy users.

2.3.2 Identity of Things

Due to rapid growth and high numbers of similar devices, reliable identification of IoT devices is an issue. The origin and history of an IoT device is especially important in security-relevant environments. In Weingaertner and Camenzind (2021), a blockchain and SSI-based system is presented that allows the manufacturer to register their devices in a nonproprietary ecosystem. During the bootstrapping process in a client environment, the device is identified by a registrar service and a connection between device and registrar is established. This connection can be documented using verifiable credentials.

This illustration demonstrates how, in the future, the identities of things will be as significant as the identities of people. In a world where objects operate on behalf of humans and it becomes increasingly unclear whether a real person or a thing is acting on their behalf, the identities of such things become increasingly crucial. The security of such devices, as well as their identity if they have one, is a major problem and part of todays' research.

2.3.3 Findy

Findy[4], a Finnish public-private organization, develops a general-purpose, shared, and secure verifiable data network. The SSI-based approach is addressing individuals as well as organizations. It is aiming to ensure the authenticity of the information required for e-services.

In 2021, the European Commission[5] proposed a framework for a European Digital Identity. The commission recommended that all EU member states work toward the implementation of the European Digital Identity framework. The solution must follow the eIDAS regulation[6]. The proposal outlines several key principles, like self-sovereign data, transparency, and interoperability. Findy is aligned with those principles and is intending to work closely with the European authorities.

2.3.4 British Columbia – Digital Government

The government of British Columbia Canada is very active in creating digital services for its citizens. Under the title "Digital Government"[7], it presents its strategy and framework toward a digital society. With OrgBook BC, it started the access to verified information about registered BC organizations in 2019. With BC Services Cards and BCeID, this path was consequently followed. The digital government platform shows how public services can enable their citizens to step into the digital world. This is done by explaining the advantages and risks as well as offering educational material.

A heated debate rages over whether a digital identity promotes inclusiveness or, on the contrary, promotes exclusion. The disadvantage of technically inept people, who are mostly elderly or uneducated, is a legitimate point.

2.4 IDENTITY TECHNOLOGY

Globally recognized standards must be created before a working system or identity layer for the World Wide Web can be built. Without these standards, only isolated solutions would develop, potentially fragmenting the identity layer and the SSI concept itself. Some technological obstacles and needs must be met in order to develop such an identity layer or ecosystem, as described in Decentralized Identity Foundation (2017):

- Independent ability to register an SSI without control or ownership by a provider.
- Being able to look up and discover identity and data across decentralized systems and cross-platform.
- Tools for users to securely store sensitive data locally and share it in a controlled manner.

Today, the standards to be defined for building these components are primarily driven by the Decentralized Identity Foundation and World Wide Web Consortium (W3C) community groups. With the definitions of Decentralized Identifiers (DID),

DID document, the secure communication via DIDComm, and verifiable credentials (VCs), both organizations have developed standards that will have a lasting impact on the SSI ecosystem. The following is an introduction to those topics to help you understand them better.

2.4.1 Decentralized Identifier (DID)

Decentralized identifiers are a standard proposed by the W3C Credentials Community Group (Sporny et al., 2021). DIDs provide a way for individuals and companies, as well as devices, to create a permanent, globally unique, and cryptographically verifiable digital identity that is completely under the control of the owner of that identity. In this context, DIDs represent a kind of address for digital identities, such as the URI (Uniform Resource Identifier) for websites. Unlike URIs, however, they must meet some additional criteria to be consistent with the SSI concept:

- No involvement of a central authority, either in registering, assigning, deactivating, or updating the associated data. Most of the today's URIs are based on DNS names or IP addresses, which are managed by a central authority.
- Control over the DID and associated metadata including their public key can be cryptographically verified. Authentication via a DID uses the same public/private key encryption scheme as blockchain technology.

The independence of the DID is intended to ensure that no central authority or company can revoke access to one's identity. Today, this is not the case, and digital identities such as e-mail addresses and social media profiles are held by the service provider. Since DIDs meet the same cryptographic criteria as blockchain technology, they also have the same characteristics in terms of security against malicious interference or tampering. The owner of the identity basically exercises control over the DID by possessing the private key associated with it.

A DID basically consists of three parts:

- the URL scheme identifier, which indicates that the object is a DID,
- the identifier of which DID method is used, and
- an address defined according to the DID method, which points to the DID document.

The individual parts are separated by a colon and a simple example of a DID looks like:

did:example:123456789abcdefghi

2.4.2 DID Document

As mentioned before, the DID points to a DID document that can be stored on a blockchain or another registry. The DID document contains the public key for the DID including information about the encryption algorithm used, other public credentials (e.g., additional public keys) that the identity holder wishes to disclose, and the corresponding network addresses. It is essential that the DID document contains no information about the owner or other person identifying information.

A DID document is typically composed of the following properties (Sporny et al., 2021):

- **Context:** Indicator for the standards used in the document.
- **DID subject:** Identifier or information about the DID that is described by the DID document.
- **Public keys:** Are required for digital signatures as well as encryption and decryption, which in turn serve as the basis for processes for authentication or the establishment of secure connections with service endpoints. If a DID document does not contain a public key, it can be assumed that the public key has been revoked or is invalid. Each public key must have an id and type property, where the id can only be used once within the DID document.
- **Authentication:** Specifies how a DID subject can cryptographically prove that it is associated with a DID.
- **Authorization and delegation:** Defines how processes can be executed on behalf of the DID subject. Delegation describes the process that a DID subject can use to authorize others to act on its behalf.
- **Service endpoints:** Can represent any type of service, such as identity management services, authentication/authorization services, social networks, data storage, etc.

2.4.3 DIDComm

Secure communication on the Internet is key. Most protocols "rely on key registries, identity providers, certificate authorities, browser or app vendors, or similarly centralized assumption" (Curren et al., 2021).

DIDComm defines a secure and private communication method that is built on decentralize identifiers. It is an agent-based approach where an agent is intended to be part of the user's wallet. The endpoints and public keys of the involved parties in the communication are obtained from the DIDs and DID documents. From a technology perspective, the DIDComm signed message is a signed JWM (JSON Web Messages) envelope.

2.4.4 Verifiable Credentials (VCs)

Credentials are an essential part of each identity system. VCs are digital credentials that are cryptographically secure, privacy respecting, and machine-verifiable (Sporny, Longley & Chadwick, 2021). Owners or holders of VCs can create verifiable presentations from them and present them to verifiers. VCs are signed by the issuer of the credential. It can be seen as a proof backed by evidence toward a claim statement of the holder. In order to verify the legitimacy of a VC, the verifier has to access a verifiable data registry, e.g., a blockchain storing the DIDs and DID documents of the issuer.

Various technology implementations for VCs exist today, but none has succeeded. Some of the buzzwords are JSON-JWT, JSON-LD, ZKP-CL, and JSON-LD ZKP with BBS+. Without convergence, there will be no functional connectedness across the ecosystems (Young, 2021).

2.4.5 Wallet

As described above, the whole concept behind SSI is based on cryptographical keys and proofs. For usability reasons, the private keys, which proof the ownership and the identity of a person or object, are stored in wallets. Those wallets can be seen as a physical wallet storing the secret information for the holder. In addition, those software wallets perform the signing process, the setup of a secure communication (DIDComm), as well as the lookup of DID and DID documents.

From the standpoint of the user, the wallet is the most used tool in the identity ecosystem. Therefore, "usability is one of the most important aspects and crucial for acceptance and adoption" (Digital Switzerland, 2022). It should be noted that today's wallet solutions fall well short of the requisite usability, and many of the standards outlined in Digital Switzerland (2022) have yet to be met.

2.4.6 Zero-knowledge Proof

One big advantage of SSI is the ability to present only selected information to the verifier. This reduces the amount of data shared and enhances data privacy. With the possibility of creating zero-knowledge proofs, this data privacy can be enhanced one step further. A zero-knowledge proof (ZKP) is a method by which one party (prover) can prove to another (verifier) that it knows information x without having to transmit it. Apart from the information that the prover knows x, the verifier learns nothing. Especially for privacy-sensitive data like person-describing data, ZKPs are an ideal way of minimizing the data footage.

In the following, a concrete example is given to clarify and describe the above concepts. Let's assume Alice has turned 18 and is allowed to visit the local casino. She (holder) holds VCs about her identity including name, birthday, residence, and body size, issued by the government (issuer). Those VCs are stored in her software wallet on her mobile phone. At the entrance to the casino, Bob wants to check Alice's age. Instead of showing her identity card with all personal information on it, she decides to use SSI. By sharing a QR code, her wallet establishes a secure communication to Bob's wallet. Both wallets act like agents. Alice's wallet creates a ZKP and transmits it to Bob's wallet. His wallet verifies the ZKP and the issuers' identity and legitimacy interacting with a node of the identity registry blockchain. Bob only gets to know that Alice is older than 18 and that this was approved by the government. With this information, nothing stands in the way of a pleasant evening and Alice is allowed to enter the casino.

2.5 GOVERNANCE

Even though SSI is based on a large technological foundation, it should not be seen as a technology-driven concept. Without a sophisticated governance framework, technology stands alone and trust in the system cannot be created.

Therefore, the Trust over IP Foundation[8] created a four-layer model with two pillars: technology and governance. The governance pillar is formed by participants of the trust ecosystem, multiple governance authorities. "A governance authority can represent any set of issuers who want to standardize the business, legal, and technical policies for issuing, holding, and verifying a set of credentials" (TrustOverIP, 2020).

Governance can be supported by technology, but nevertheless it is grounded on agreements, policies, rules, and trustworthy participants. Governance on the first layer – the utility frameworks – is ensured by node operators or stewards maintaining a blockchain or DLT infrastructure to allow credential issuers to store proofs on this infrastructure. On layer 2 – the agents' frameworks – the role of governance authorities is to verify and certify software, hardware, and cloud providers. Since wallets have a key role to play, their integrity and conformity to privacy, security, and data protection standards is crucial. Layer 3 – the credential frameworks – assures that trusted authorities put their influence behind credentials and issuers to meet the standards and to protect the holders. During this quality assurance process, sanctions may also be necessary. Finally, on the top layer (4) – the ecosystem frameworks – one governance authority can oversee the whole ecosystem to ensure interoperability and audit compliance. Each ecosystem stands for a set of identities and certificates. Examples for such ecosystems are universities with their certificates for degrees, governments with various certificates about citizens' identities, or road traffic authorities with certificates for driver licenses.

Trust is the foundation of the entire ecosystem. On the one side, there is technological trust, in which one trusts the cryptographic procedures utilized. On the other hand, it is human trust in organizations and the governing process as a whole. This is an important component of the trust ecosystem. One must believe that credential issuers are who they say they are, and that the system will remain stable and resilient no matter what happens. Through decentralization, blockchain may play a key role in this ecosystem by ensuring immutability and stability.

2.6 CRITICAL ASPECTS OF SELF-SOVEREIGN IDENTITY

SSI is a relatively new concept. This is the most serious critique. Many experts remain skeptical since there is no long-term experience of working systems built with this technology. Another important point that is frequently cited and is true is that the user has complete control over her or his data. On the one hand, this is one of the most significant benefits, as it leads to censorship-free usage. However, this is a significant duty, and many people question that the majority of users are capable. It is essential that deputy regulations and emergency remedies be implemented, but they can also be utilized for paternalism. One thing is certain: we will not know until we implement the technology in a real-world use case at a wider scale. As a result, many experts support the deployment of real use cases as well as further research. Furthermore, governance, inclusivity, usability, and even ethical issues can only be investigated as part of a larger ecosystem.

2.7 CONCLUSION

Digital identity is one of the major infrastructure services needed for the digital world. Many use cases rely on a trustworthy, reliable, and available digital identity. One important aspect of such a digital identity is the independence from single organizations or authority bodies. SSI offers a framework of technology components and governance guidelines to build such a transparent, independent, privacy protecting, and interoperable identity infrastructure where the user is in control of their data. Blockchain is intended to play the role of a trust anchor in this infrastructure. Its immutable and distributed character ideally supports the concepts of SSI.

Today, we see a large rise in the number of local and countrywide initiatives driving SSI in their ecosystems. A crucial aspect for the flourishing and durability of those initiatives and solutions will be their interoperability. The digital world does not end at a country border or a business process. The World Wide Web is a world-spanning infrastructure supporting all kinds of use cases and allowing unrestricted access for everybody. With the growing number of IoT devices, this event exponentially rises.

NOTES

1. https://www.merriam-webster.com/dictionary/identity
2. www.cardossier.ch
3. https://youtu.be/kBUD3kS6rj0
4. www.findy.fi
5. https://eur-lex.europa.eu/eli/reco/2021/946/oj
6. https://eur-lex.europa.eu/eli/reg/2014/910/oj
7. https://digital.gov.bc.ca/
8. www.trustoverip.org

REFERENCES

Allen, C. (2016, April 25). *The Path to Self-Sovereign Identities.* Life With Alacrity. Retrieved December 6, 2021, from http://www.lifewithalacrity.com/2016/04/the-path-to-self-soverereign-identity.html

Curren, S., Looker, T., & Terbu, O. (2021). *DIDComm Messaging.* Retrieved December 2, 2021, from https://identity.foundation/didcomm-messaging/spec/

Decentralized Identity Foundation. (2017, October 11). *The Rising Tide of Decentralized Identity.* Retrieved December 10, 2021, from https://medium.com/decentralized-identity/the-rising-tide-of-decentralized-identity-2e163e4ec663

Digital Switzerland. (2022, April). *Building a Swiss Digital Trust Ecosystem - Perspectives Around an e-ID Ecosystem in Switzerland.* Retrieved April 2022 from https://digitalswitzerland.com/wp-content/uploads/2020/04/Building-a-Swiss-Digital-Trust-Ecosystem_digitalswitzerland_April2022_vF.pdf

Liu, Y., Zhao, Z., Guo, G., Wang, X., Tan, Z., & Wang, S. (2017, August). An identity management system based on blockchain. In *2017 15th Annual Conference on Privacy, Security and Trust (PST)* (pp. 44–56). New York: IEEE.

Microsoft Corporation. (2018). *Decentralized Identity. Own and Control Your Identity.* Retrieved December 10, 2021, from https://query.prod.cms.rt.microsoft.com/cms/api/am/binary/RE2DjfY

Mühle, A., Grüner, A., Gayvoronskaya, T., & Meinel, C. (2018). A survey on essential components of a self-sovereign identity. *Computer Science Review*, 30, 80–86.

Sovrin Foundation. (2018, January). *Sovrin: A Protocol and Token for Self-Sovereign Identity & Decentralized Trust*. Retrieved December 10, 2021, from https://sovrin.org/wp-content/uploads/Sovrin-Protocol-and-Token-White-Paper.pdf

Sporny, M., Longley, D., & Chadwick, D. (2021, November 9). *Verifiable Credentials Data Model v1.1 - Expressing Verifiable information on the Web*. Retrieved December 2, 2021, from https://www.w3.org/TR/vc-data-model/

Sporny, M., Longley, D., Sabadello, M., Reed, D., Steele, O, & Allen, C. (2021, August 3). *Decentralized Identifiers (DIDs) v1.0 - Core Architecture, Data Model, and Representations*. Retrieved December 2, 2021, from https://w3c.github.io/did-core/

Stokkink, Q., & Pouwelse, J. (2018, July). Deployment of a blockchain-based self-sovereign identity. In *2018 IEEE International Conference on Internet of Things (iThings) and IEEE Green Computing and Communications (GreenCom) and IEEE Cyber, Physical and Social Computing (CPSCom) and IEEE Smart Data (SmartData)* (pp. 1336–1342). New York: IEEE.

TrustOverIP. (2020, May 5). *Introducing the Trust Over IP Foundation*. Retrieved December 7, 2021, from https://trustoverip.org/wp-content/uploads/2020/05/toip_introduction_050520.pdf

van Bokkem, D., Hageman, R., Koning, G., Nguyen, L., & Zarin, N. (2019). Self-sovereign identity solutions: The necessity of blockchain technology. arXiv preprint arXiv:1904.12816. Retrieved April 29, 2019, from https://arxiv.org/abs/1904.12816

Weingaertner, T., & Camenzind, O. (2021). Identity of Things: Applying concepts from Self Sovereign Identity to IoT devices. *The Journal of British Blockchain Association*, 4, 1–7.

Young, K. (2021, February). *Verifiable Credentials Flavors Explained*. Retrieved May 20, 2022, from https://www.lfph.io/wp-content/uploads/2021/02/Verifiable-Credentials-Flavors-Explained.pdf

Section II

Privacy Dilemma

3 Security and Corporate Violation to Privacy in the Internet of Things Age

Darine Amayed
École de technologie supérieure, Montréal, Québec, Canada

Fehmi Jaafar
Department of Mathematics and Computer Science,
Québec University at Chicoutimi, Québec, Canada

Riadh Ben Chaabene
École de technologie supérieure, Montréal, Québec, Canada

Mohamed Cheriet
École de technologie supérieure, Montréal, Québec, Canada

CONTENTS

3.1 Introduction 40
3.1.1 Privacy Issues Revisited 41
3.1.2 Related Surveys 41
3.1.3 Chapter Structure 42
3.2 Internet of Things (IoT): Evolution 42
3.2.1 History 43
3.2.1.1 Radiofrequency Identification 44
3.2.1.2 Wi-Fi 45
3.2.1.3 Bluetooth 46
3.2.2 Present: The Development for Distributed Networking and Ubiquitous Computing 46
3.2.2.1 Cloud Computing 46
3.2.2.2 Edge Computing 46
3.2.3 Future 47
3.2.3.1 Artificial Intelligence (AI) 48
3.2.3.2 5G 48
3.3 Application Areas of IoT 49
3.3.1 Applications 49
3.3.1.1 Smart Home 49
3.3.1.2 Smart Cities 50

DOI: 10.1201/9781003227656-6

3.3.1.3 Healthcare 50
3.3.1.4 Retails 50
3.3.1.5 Transportation 51
3.4 Privacy in the Digital Era 51
3.4.1 Anonymity 52
3.4.2 Control and Fairness over Personal Data 52
3.4.3 Confidentiality 53
3.4.4 Conformity to Laws and Policies 53
3.4.4.1 2018: The General Data Protection Regulation (GDPR) 53
3.4.4.2 2020: The California Consumer Privacy Act (CCPA) 54
3.5 IoT Security Issues to Privacy 54
3.5.1 Technologies to Enhance Security Issues 56
3.5.1.1 Blockchain 56
3.5.1.2 InterPlanetary File System 58
3.5.1.3 Artificial Intelligence 58
3.5.1.4 Fog Computing 60
3.6 Corporate Violation 60
3.7 Conclusion Future Work 63
References 64

3.1 INTRODUCTION

One of the objectives of smart spaces and systems is improving society and human life quality, either in terms of comfort, efficiency, or productivity. The Internet of Things (IoT) has developed into a backbone innovation for smart environments capable of providing the mentioned features. In a market research study, the report evaluates that IoT systems could reach a potential economic impact of $11 trillion in the year 2025 compared to the actual $4 trillion, which represents approximately 11% of the world economy Institute [33]. IoT systems are becoming one of the most developed fields in computer science [74], and it's a significant contributor to the world economy. Their real-time data processing and analysis greatly benefit multiple fields and infrastructures, especially for healthcare systems, transport, and autonomous vehicle technology, smart cities, etc. Privacy and security are critical elements in any real-world smart environment established by an IoT system. All of this made them a target of cyberattacks due to their security leak compared to conventional computers. The report has shown an increase in attacks targeting IoT devices by 600% in the year 2017 Symantic [65]. IoT devices, despite their importance, lack major security updates, have insecure interfaces, and have insufficient data protection. For example, the Mirai Botnet, first introduced in 2016, was able to target almost 400,000 devices simultaneously, leading to a total of $100 million in damage and shutting down significant platforms such as Netflix, GitHub, and Reddit for multiple hours [48]. IoT devices don't have the same or enough computational and hardware resources to use complex security features such as computers or smartphones [2]. Those systems can use a security layer in their implementation that provides the required security against most cyberattacks, such as firewalls, antivirus software, etc. Hence, it is crucial to focus on securing those systems. They are becoming increasingly targeted by cyberattacks due to their security vulnerabilities. Moreover, companies are using IoT to access users' data to create profiling that will help target them for personal reasons. This creates a breach of their right to privacy. In this era, Big Data represent financial benefits; that is why we see their tremendous growth since every organization tries to incorporate them in their business model for their benefit to collect an enormous amount of data.

3.1.1 Privacy Issues Revisited

As the Internet is taking over our daily lives, most of the world's data, such as photos, videos, social information, and many more, is being constantly published and shared over the global network; thus, privacy is becoming a major target and a right that keep fading over time which raises prime concerns [12]. Moreover, the rise of IoT devices connected to the Internet and collecting extensive data carries a comprehensive risk to users' privacy. An entity has the right to determine the amount of data it is willing to provide with a system or others. Still, despite this, many labeled data of a person can be collected without the person's consciousness when using IoT devices. Having control of personal data is becoming difficult, if not impossible, especially when, in multiple cases, that information is necessary for the device's proper function. Especially the identification of an individual and his behavioral patterns is a growing concern. As IoT devices are increasingly used in all fields of daily life, such as in the healthcare sector [40], a great amount of commonly considered private information is collected by sensors and stored for analyses and study reasons, all potentially without adequate accountability, transparency, security, or meaningful consent. It is clear to say that consumers are surrendering their privacy without noticing. As a wearable, smart applications, and Wi-Fi increase demand and replace "old" devices. Consumers won't be capable of purchasing devices that won't collect their data and track them. So, it becomes necessary to discuss the implications of the IoT and the need to integrate privacy principles and safeguards into creating and implementing smart environment components.

3.1.2 Related Surveys

There are various existing surveys on IoT security and privacy issues. Hassija et al. [30] have summarized various security threats that affect IoT systems and techniques to use to enhance security. Weber [73] has discussed privacy concerns and then the need for new requirements to deal with IoT breaches of privacy. He describes legal approaches for protecting privacy that needs to be developed. Authors

Chabridon et al. [12] focused mainly on IoT layers' security issues and provided their security issues. Khanna and Kaur [41] have researched the application of the IoT, their issues, and their challenges. Sayar [60] presented the rise of the General Data Protection Regulation as a privacy regime that impacts IoT privacy issues. In Sharma et al. [63], the authors summarized the history of the IoT to its future. Mohammad.S et al. in Mahdavinejad et al. [48] have discussed the relationship between IoT and machine learning, and in Huang et al. [31], the authors described the relation between IoT and blockchain.

3.1.3 Chapter Structure

The chapter's organization is as follows: Section 3.2 describes the evolution of the IoT, how it started, and how it did evolve over the years. Section 3.4 discusses how digitization affects privacy and the recent policies created to regulate data privacy. Section 3.3 summarizes the architecture used in the IoT and its significant application. Section 3.5 shows security issues that IoT is facing and how some of the new technologies could help avoid them. Section 3.6 is where we go in-depth into how our privacy is being violated by companies using IoT and the measures that we need to use to avoid losing it. Initially, in Section 3.7, we conduct our conclusion and provide insight for future studies.

3.2 INTERNET OF THINGS (IoT): EVOLUTION

The IoT is a massive network of connected things and people. They all collect and share data about their application and their surrounding environment. It's a concept of devices connected to the Internet and other devices. They typically range from small sensors to complex controllers offering all types of monitoring and control services to amplify and automate daily tasks [61].

An IoT ecosystem embrace web-enabled smart devices that utilize embedded systems, like sensors, communication hardware, and processors, to collect, send and use the data they obtain from their environment [45]. They share the collected data by an IoT gateway or other edge devices, which are analyzed either in a cloud system or locally. Moreover, those devices can communicate with each other to perform a collaborative act by using their collective information. Human interaction is minimal with IoT systems; they either set up the system, access their data, or give instructions, but their functioning is mostly automated. Figure 3.1 represents an example of an IoT system.

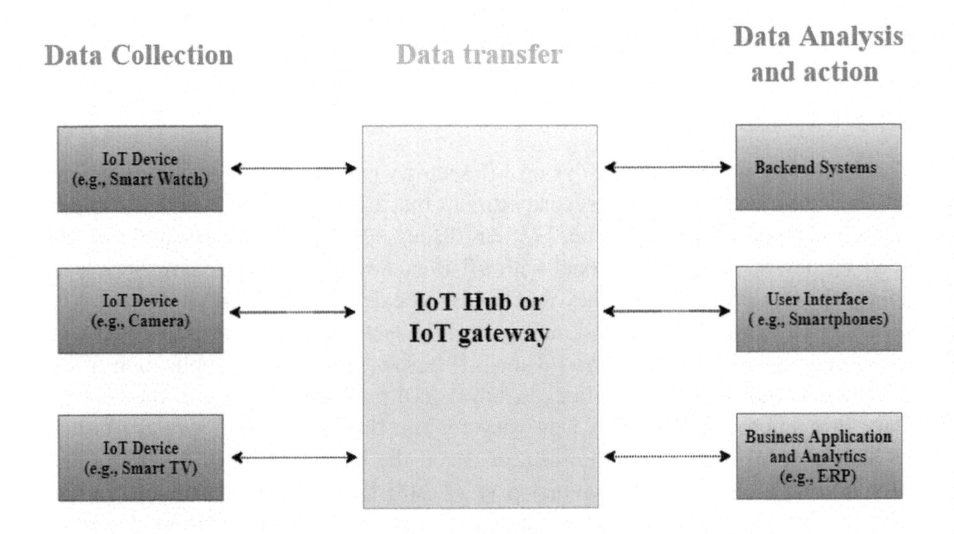

FIGURE 3.1 Example of an Internet of Things system.

3.2.1 History

The term IoT was first documented by Kevin Ashton in MIT Auto-ID center in 1999 [7]. It was first designated to describe systems that could connect and communicate with the Internet via a universal network of data sensors. This description has changed over time beyond the original intention. But, it is essential to study the root of it all to understand its full context. From the birth of the Internet in 1989, the idea of connecting things on the Internet has grown tremendously, leading to the creation of a massive field of research. One of the first creations in this field was a coffee toaster created by John Romkey in 1990 [57]; it was able to be turned on/off using the Internet. 1994 saw the birth of the first wear cam created by Steve Mann, with almost real-time performance. By 1997, sensors started to get global attention about their importance in the future and their course to achieve automation tasks, and this was explained by Paul Saffo in his description of sensors and their future course [59]. All of this led the way for Kevin Ashton to create the term IoT during a presentation for Procter Gamble, where he described IoT as a technology that connected several devices with the help of radiofrequency identification (RFID) tags for supply chain management. In early 2000, the Internet was thriving, and everyone wanted to contribute; it started with LG presenting the first-ever smart refrigerator that would automatically determine whether or not food items were replenished. By that time, it was clear that most industries were training to adapt to this new technology. RFID started to be deployed at an enormous rate. In 2003 retail giant Walmart deployed RFID in all its shops globally, and soon multiple companies followed that path. In 2008, a group of companies launched the IPSO Alliance to promote Internet Protocol (IP) use in networks of " smart objects" and enable the IoT. And by the end of 2011, IPv6 was introduced, leading to a massive interest in this field, and IoT was adopted by major communication companies such as IBM, Cisco, and Google. Figure 3.2 presents a full schema of the evolution of IoT from 1990 to 2016. Three major technologies responsible for the breakthrough of IoT are discussed below, and Table 3.1 summarizes the main differences between them.

TABLE 3.1
Comparison between RFID & Bluetooth & Wi-Fi

	RFID	Bluetooth	Wi-Fi
Communication	Unidirectional (one-way)	Bidirectional (two-way)	Bidirectional (2-way)
Range Coverage	3 meter	100 meter	100 meter
Operation Frequency	Varies	2.4 GHz	2.4 GHz–5 GHz
Set up Time	¡0.1s	¡0.1s	¡6s
Network Type	Point-to-Point	Point-to-Point	WPAN
Data Rate	Varies	22Mbps	144Mbps
Security	Hardware/Protocol level	Protocol level	Hardware/ Protocol level
Application	EZ-Pass, Tracking items	Communication with peripherals	Wireless Internet

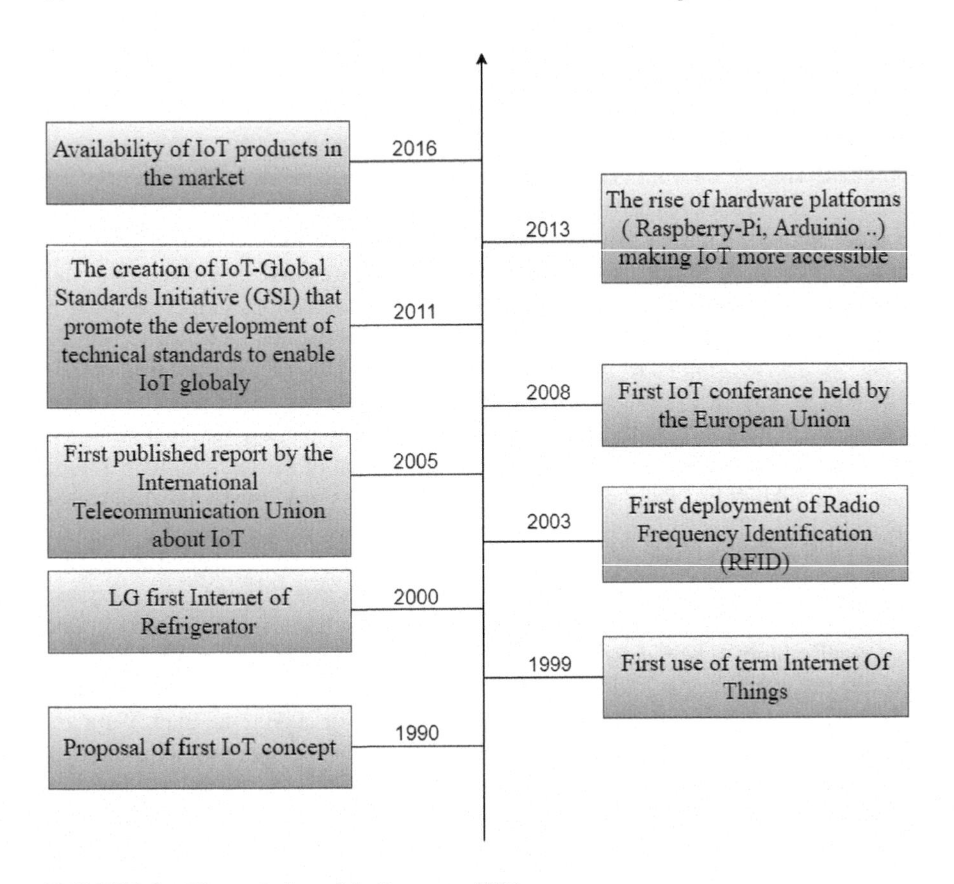

FIGURE 3.2 The evolution of the Internet of Things.

3.2.1.1 Radiofrequency Identification

By connecting smart devices, the world has become more accessible. Smart devices are being used everywhere, making IoT popular with businesses and services across all industries. One of those technologies is RFID. As mentioned before, the term "Internet of Things" was introduced during a presentation about RFID. RFID is a technology that uses radiofrequencies to transmit data [39]. It automatically identifies an object and captures its data stored in a small microchip tag. This process is performed using RFID tags, they come in many different shapes, sizes, and capabilities. These tags are of two kinds.

- Active tags: contain internal power sources such as battery power. They can exchange communication with other tags and could automatically communicate with the RFID reader.
- Passive tags: They are powered by the tag reader. Hence, they don't require any internal power.

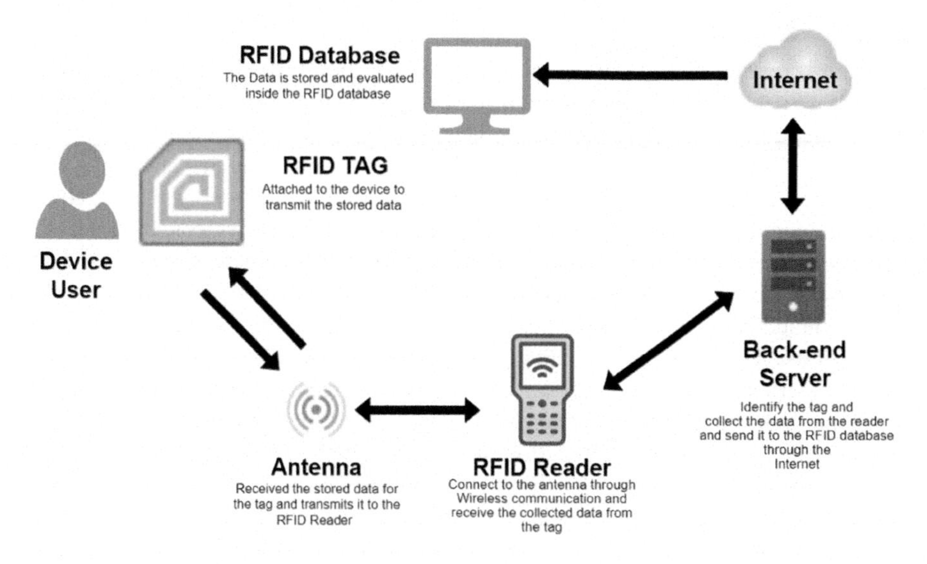

FIGURE 3.3 Radiofrequency identification system.

Their main objective is to store the data and the object's identity. On the other hand, the RFID reader communicates with the RFID tag by radio waves to read their data remotely. The data is then transferred from the scanning device to the central company server, where the data is stored and analyzed. The function process of RFID is represented in Figure 3.3.

Data capturing and automated identification represent the main advantages of RFID technology, Islam et al. [36] opened a new perspective to business activities by reducing the cost of their similar previously used systems such as barcodes. RFID technology provided excellent potential for IoT thanks to their usage and especially their different size; it enabled their deployment in various areas of the environment. Moreover, with the RFID reader being able to communicate throw Internet terminals, it becomes possible to automatically and in real-time identify, track, and monitor the objects attached with tags at a global scale [44].

3.2.1.2 Wi-Fi

Another major contributor to the IoT is Wi-Fi, a globally accepted wireless communication used to send/receive data, signals, commands, and more providing low cost of installation and maintenance since it only requires a transceiver. Despite that, it offers multiple advantages compared to other wireless technologies; it can transmit at frequencies of 2.4 GHz or 5 GHz, meaning that signals can carry more data. Moreover, its data transfer rate can reach 300Mbps data transfer rate and about 100M to 150Mbps throughput, which is massive for the growth and performance of the IoT since it contributes to minor delay and better real-time [49].

3.2.1.3 Bluetooth

Since its creation in 1998, Bluetooth has seen and provided considerable technological advancement. Its contribution to device-to-device communication has led to its adaption by multiple industries. It is no coincidence that his technology is widespread and found in most devices such as phones, smartwatches, PCs, glasses, earphones, shoes, and many more. So, it is safe to say that it represents the go-to wireless connectivity solution for wearable gadgets and other devices. It uses UHF radio waves, known as short wave radio, with radio bands ranging from 2.402 GHz to 2.480 GHz and building a Personal Area Network (PAN), enabling a master-slave connectivity [58]. This technology provided the reality of IoT and its wide range of usage. And with the adaptation of Bluetooth low energy (BLE) that can be found in battery-powered devices, mostly sensor devices that offer reduced power consumption, reduced latency, and cost compared to the classic Bluetooth while maintaining a decent communication range, more opportunities and benefits were provided to the field of IoT [62]. Since BLE can help conserve the device's energy when it is not in use, it can quickly pair and reconnect with devices in less than six milliseconds. With Bluetooth product selling crossing 4 billion this year and its advantages, experts believe that Bluetooth will be a significant factor in the trillion-dollar IoT market.

3.2.2 Present: The Development for Distributed Networking and Ubiquitous Computing

Previously discussed technologies conducted the growth of IoT in today's era. But, with technological advancement and especially the development of the world data reaching around 64.2 zettabytes in 2020, a new solution needed to be adapted to provide more efficiency and performance to the IoT field.

3.2.2.1 Cloud Computing

Due to the rapid growth of technology, Storage issues, processing, and analyzing large amounts of data, cloud computing is becoming essential to IoT. Cloud computing enables users to perform computing tasks using services provided over the Internet [19]. The "Cloud" is a centralized system that helps in transferring and delivering data to data centers over the Internet, also making access to a large amount of data and programs accessible and fast. Moreover, Cloud Computing allows economic solutions since it doesn't require on-site storage, processing, and analytics infrastructure. And as we said, we live in a significant data era nowadays, and scalability is a must. Cloud computing can provide that, meaning with business and data growth, technological, and analytical capabilities can also. Researchers saw an opportunity in the cloud for IoT systems. Their mutual combination enabled a vast amount of data collection by the IoT devices and powerful processing for those data streams and their monitoring using the cloud [20].

3.2.2.2 Edge Computing

It is a distributed information technology architecture in which collected data is processed at the network boundary, as close to the originating source as possible.

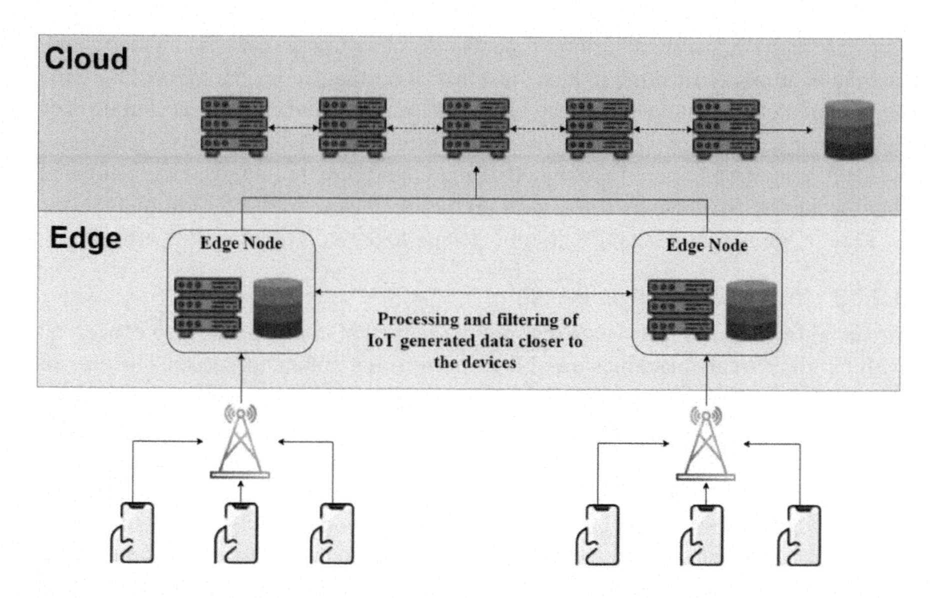

FIGURE 3.4 Edge computing architecture.

Traditional computing processes rely on a centralized data center and the amount of data every second counts. Latency issues, bandwidth limitations, and unpredictable network disruptions can be considered since they could provide ambiguous scenarios. An efficient way to respond to those challenges is edge computing architecture [28]. Edge computing moves some portion of the storage and computes resources out of the central data center and closer to the data source itself, as demonstrated in Figure 3.4. Rather than transmitting raw data to a central data center for processing and analysis, that work is performed where the data is generated. It this well suited for IoT systems. It enables the data collected by sensors to be gathered and processed at the device location rather than sending it back to a data center or cloud. The need for IoT devices to consume compute power is increasingly valuable for real-time analysis. Moreover, having access to a close computational power can reduce a large number of issues, such as the latency of communication between IoT devices and the central IT networks those devices are connected to, the slower response time when sending data back-and-forth between the device and the cloud and, network bandwidth issues when required to send large amounts of data over slow cellular or satellite connections [10].

It enhances rapid decision-making by deploying analytics algorithms and machine learning models processing locally.

3.2.3 Future

The future of IoT devices is certain; they will only keep growing. It has the potential to be limitless. Advances to the industrial Internet will only accelerate the need and usage of IoT devices through increased network agility, integrated artificial intelligence (AI), and other technological advancements. And seeing that those devices are just starting to dominate a variety of fields, it is clear to say that it won't stop them, and the numbers confirming indicate just that: By 2022, Google Home will have the largest IoT devices market share, at 48%. The average number of connected devices per household in 2020 was 10. In 2021, 35 billion IoT devices were installed worldwide, making the number of connected devices in 2021 46 billion. And it is estimated that the number will increase to almost 80 billion devices by 2025 Columbus [14].

Like in the past and the present, more technologies contribute to this growth.

3.2.3.1 Artificial Intelligence (AI)

Artificial Intelligence (AI) is a wide-ranging branch of computer science concerned with building smart machines capable of performing automated tasks. The area of AI has attracted significant interest over recent years. Many domains are using it for their development. But to perform efficiently requires a large volume of data that IoT devices could acquire. Therefore, combining both technologies can unlock more significant potential. When AI is added to the IoT, it allows those devices to analyze data, make decisions, and act on it without human involvement by humans [43].

Machine learning, an AI technology, enables creation models to automatically identify patterns and detect anomalies in the data that smart sensors generate, such as temperature, pressure, humidity, air quality, vibration, and sound. Traditionally those tasks require human intervention; now, they happen instantly and are automatically operational, which enables predictions up to 20 times earlier and with greater accuracy. Also, Other AI technologies such as computer vision, speech recognition, and text mining could extract insight from data providing new execution areas for IoT [48].

AI applications for IoT enable companies to avoid unplanned downtime, increase operating efficiency, spawn new products and services, and enhance risk management. Moreover, AI-powered IoT can improve operational efficiency. It can predict component failure driving conditions and identify parameters to be adjusted immediately to maintain ideal outcomes by detecting patterns invisible to the human eye.

3.2.3.2 5G

This network presents the fifth-generation wireless technology. It can provide higher speed, lower latency, and greater capacity than previous LTE networks. It benefits IoT devices with ultra-low latency and ultra-fast response times. The 4G network response is 400× faster than its processor, with predicted speeds of up to 10 Gbps. Furthermore, 5G networks will have even lower latency than 4G LTE, with data transmission taking less than five milliseconds. Finally, its capacity to connect more devices at once is considered to deliver up to 1,000× more power than 4G, creating a productive ground for IoT development [46]. It will become possible to involve thousands of sensors in hundreds of devices to act together and operate automatically. For example, Smart homes and cities will make a massive step into the future with 5G; by using multiple connected devices, AI will be taken to places it has never been before with edge computing. From houses that give personalized suggestions that maximize environmental impact to connected vehicles for police linked to traffic lights. This high-speed, high-capacity, low-latency application of 5G will impact each sector. As a result, it will create an excellent and massive IoT ecosystem where networks can serve billions of connected devices [13].

3.3 APPLICATION AREAS OF IoT

IoT technology has a wide variety of applications, and its use is growing tremendously. Among other things, the IoT has evolved into an enabler for intelligent devices used in many aspects of our society, including our homes, hospitals, and office buildings. That said, the fast development of the IoT amid growing cybersecurity problems has led to widespread concern regarding a world of interconnected devices and user privacy.

Despite all of this, those layers possess security issues and have been targeted by cyberattacks. Table 3.2 shows each layer attack.

3.3.1 Applications

The IoT is not a mysterious hype anymore. It starts shaping our future. IoT results from humans seeking comfort and adaptation to new technological advancements. That's why the creation of smart devices is constantly growing. With data being the new currency and tons of crucial concerns can be addressed and resolved through data, IoT is being implemented in almost every imaginable field [6].

3.3.1.1 Smart Home

The first generation of smart homes was more about remote control and automation. Years ago, an automated task such as operating a blinder via your smartphone or making your thermostat remember the temperature you prefer was acceptable to call a smart home [5]. In 2021 with IoT, we are way far from that. IoT blends intelligent utility systems and entertainment by connecting all the devices. In smart homes, various sensors are deployed, which provide intelligent and automated services to the user. They tend to create an automated routine for our daily tasks. They learn about our habits and determine consumption patterns like energy conservation by automatically turning off lights and electronic gadgets or everyday water usage [1]. For example, the project, Cook et al. [15] provide daily triggered automated tasks in response to the user routine. This made using intelligent agent communication together. Moreover, it provides multiple security options, such as advanced locking systems connected to surveillance systems.

TABLE 3.2
IoT Architecture Layer Common Cyberattacks

	Perception Layer	Network Layer	Processing Layer
Cyberattacks	• Node Capture Attacks • Malicious Code Injection • False Data Injection • Eavesdropping • Booting Vulnerabilites	• Phishing Attack Ahmad et al. • Access Attack Burhan et al. • DOS/DDOS Attack • Routing Attack • Unlwaful Attacks	• Cloud Flooding Attack • Cloud Malware Tewari and Injection • SQL Injection • Men-In-The-Middle Attack

Smart home IoT systems are very beneficial for the elderly. Their daily condition and health are monitored using floor sensors that track their movement across the house. They are directly informed in an emergency and help detect if someone falls.

3.3.1.2 Smart Cities

The idea is to ensure comfort, savings, and security in the city as much as possible. Making everything connected is a good way to transmit information as quickly as possible and make decision-making efficient and instantly. This is why IoT is a major contributor to the creation of smart cities as stated by Alavi et al. [3] and Qian et al. [56]. In fact, IoT can provide the following services:

- Traffic control: Receiving data from sensors and cars to adjust traffic lights in real-time reduces road congestion.
- Garbage collection: By creating scheduled pick-up in need compared to a preplanned schedule by the distribution of smart garbage cans across the city that can send data to waste management companies.
- E-governance: Making a mobile driver's license and ID card with digital credentials speeds and simplifies access to the city and local government services.

Barcelona Gascó-Hernandez [24] is considered one of the world's most advanced smart cities. It has a CityOS project aiming to create a single operating system for all smart devices and services in the city. It has sensors all over the city, for example, water storage tanks and water supply lines, enabling it to predict domestic and industrial usage water requirements.

3.3.1.3 Healthcare

Monitoring a person's health condition using wearable devices benefits the health sector. Many wearable devices are being created using different types of sensors to keep track of personal health conditions and provide warning alerts in case they detect an anomaly or abnormal indicator; they could even suggest prescriptions to their user [55].

In an experiment by Sysoev et al. [66], they used a stress recognition application using smartphone sensors to measure the stress level of a college student. They understood college student stress and their academic performance by tracking the student's location during a full day via GPA sensor, human interaction via audio sensors, physical activities, and sleep and rest amount.

IoT applications can also create a record of all the person's medical details called an Electronic Health Record (EHR). It could contain blood pressure, allergies, sugar in the blood, and many more [18].

3.3.1.4 Retails

Like many industries, retail is embracing IoT transformation. From revenue growth, customer experience, cost reductions, and process optimizations, smart retail is changing the business [72].

IoT sensors connected to a dashboard, like color-coded buttons or emotional analysis sensors, allow stores to collect customer feedback immediately after the shopping experience. The provided data can help the store to improve its customer experience. Tracking sensors are used all over the store to monitor their goods or assets. The retailers can now track the location, humidity, temperature, and stock to ensure higher quality control and ensure food doesn't spoil. Also, they help determine if the equipment is safe, delivered on time, and transported in ideal conditions. They are also tracking shopping carts and baskets, helping to reduce the cost of replacing them. Other examples of IoT in retail are smart shelves. A lot of wasted time and energy is used to avoid items out-of-stock or misplaced. That task could be automated using smart shelves. Using weight sensors and RFID tags, they could inform employees when items are running low or incorrectly placed on a shelf. This saves time and human error and even provides security by detecting potential theft [11].

Moreover, the most recent example is the usage of counting systems. Due to the COVID-19 pandemic and government regulations. Those devices help track the number of people entering or exiting in real time and provide alerts when the capacity threshold has been met.

3.3.1.5 Transportation

Transportation/Mobility is the second largest IoT application area in 2020. The global smart transportation market could reach $262 billion by 2025, thanks to the value of IoT in vehicles. However, the benefits don't end with financial success for car manufacturers. It's improving nearly every aspect of the industry. Tesla set the industry benchmark for connected cars when it launched the Model S in 2012, introducing the first over-the-air software update capabilities. Since then, every other company has tried to copy that formula. With IoT communication, vehicles would have real-time data on everything else on the road, making driving safer. This connectivity is also a factor in fully automated cars [37]. With the vehicle being able to connect to others and its environment, it could navigate more effectively. Without IoT, safe and reliable self-driving cars wouldn't be possible. Using IoT, it becomes possible to identify potential problems before they become expensive. Additionally, they can decrease fuel consumption through more efficient routes, better-driving behavior, and scheduling optimization.

3.4 PRIVACY IN THE DIGITAL ERA

It's becoming clearer that data is the fuel for technological advancement. Billions of devices, sensors, and cameras are gathering data each millisecond. And most of those data are generated by humans and their personal information. Our need to control what we hide and share is no longer in our hands. Privacy is becoming a significant concern in this technological era. According to a poll conducted in June 2019, about 74% of Internet users in the United States are more concerned with their online privacy than ever, but only 33% of Internet users in the United States are aware of their country's privacy and data protection rules. And some of them reported that they don't distinguish between what represents private and public data.

Privacy's definition varies from one person to another depending on the thing and the context. This section will examine several cores of digital privacy that should carry over people and their interactions on the Internet.

3.4.1 Anonymity

Anonymity can be qualified as a condition of avoiding identification. It's essential to the realization of human rights and fundamental [50].

Imagine walking in a mall covered with signs all over your body. Those signs indicate every store, the store you visited, the things you bought, what you like, what the things you looked at, and so on. This varies close to what could happen when connected to the Internet.

Surfing the World Wide Web, we think that we have a confident expectation of anonymity, like in the real world, we don't expect that individuals are tracking us or trying to observe us. But unfortunately, this is not the case. Using the Internet or anything connected to it generates endless trails of data specifying every step we make. Especially now with the growth of social media and digital advertising, "haven't we all said we are scrolling down on our social media," "How did I get an ad for this device that I just thought about?" Every move we make is registered and used by advertising companies to target us. With data being everywhere, either transactional data, mouse dropping, streaming data, or media data, we can create an online life profile of each individual. A technology that made all of this possible is "cookies" [17]. Stored directly on our hard drive, they provide the ability to connect to the website or web servers, collect information related to our online activities, and save it for future usage. Intended for the harmless motivation behind empowering Web sites to perceive a recent visitor and react appropriately, cookies were immediately taken on by Web locales and giant companies to target explicit personal activities for advertising. Companies such as Meta, Amazon, Netflix, and almost most web-based companies collect personal information to create fully identifiable profiles on a user's online and offline behavior. Moreover, the rise of identification technologies, such as fingerprints, voice, facial recognition, and many other types of use of biometric data, is wide-spreading, leading to zero anonymity.

3.4.2 Control and Fairness over Personal Data

Personal data, also known as Personally Identifiable Information (PII) ONIK et al.

Onik et al. [52] means any information which can be used to distinguish or trace the identity of an individual (e.g., name, social security number, biometric records, etc.) alone, or when combined with other personal or identifying information which is linked or linkable to a specific individual (e.g., date and place of birth, mother's maiden name, etc.). The user's right is to know what data is being collected, who is using it, and for what purpose.

Again, imagine going to a merchant, a doctor, or a bank. You anticipate that those experts/organizations will gather the data gathered concerning the service provided and use it for the sole motivation behind offering the requested service. The merchant will use it to process the statement and the shipping, the doctor for your health, and the bank to manage your account, and it ends there. Alas, today's online or offline practices don't consider this role of privacy. Any information left behind will be gathered and used for alternative purposes without the user's knowledge or consent. We tend to share everything over the Internet; since the world is going digital, even standard services remove the physical aspect and turn it into only a digital one. Name, date of birth, credit card number, address, list of transactions, medical record, appointments, and everything you share over the Internet are stored somewhere and used for a person or a company's purpose. There are many examples of companies using and disclosing personal data well beyond what the individual intended. One of the famous Isaak and Hanna [35] which raised the alarm about data privacy was obtaining a political data-analytics firm named Cambridge Analytica personal data from over 87 million Facebook users to assist the 2016 presidential campaigns of Ted Cruz and Donald Trump.

3.4.3 Confidentiality

If personal data become public, confidentiality and privacy are lost [23]. We expect it to be read only by the intended recipient when we send a virtual message. This expectation is no longer safe. A virtual message is just one example. Today, our communications, medical records, and confidential documents are on the Internet, creating a high risk and enormous consequences for our privacy. If an unauthorized party gets access to that information, you become an easy target, either for criminal actions or blackmailing, or someone could delete it.

3.4.4 Conformity to Laws and Policies

In the digital world, privacy must be seen as a crucially important right for us as a society. At the conceptual level, it must go through the same process of evolution as its older sibling, the right to freedom of expression. Clearly, the existing legal framework needs to adapt to those new concerns; otherwise, privacy will be a nonexisting thing with the enormous collection of data and the growth of the latest technology. That way, multiple new legalizations, and laws are beginning to readjust the terms of privacy and what is legal and not during this digital age. Below are some of the new rules that are conducting this change.

3.4.4.1 2018: The General Data Protection Regulation (GDPR)

The General Data Protection Regulation (GDPR) [70] is a law dealing with data protection and privacy that went into effect in the European Union (EU) and the European Economic Area EEA) on May 25, 2018. It also applies to the transfer of personal data outside of the EU and EEA. Now, more related to personal data security, we present the seven principles of the GDPR that represent the rules that each company or individual needs to follow in dealing with people's customers' personal data [69]. The principals are as follows [71]:

- Lawfulness, fairness, and transparency: The term lawfulness refers to the reasons for personal processing data, meaning that every piece of information related to data usage should be well-defined and specific.

Fairness and transparency go hand and hand with lawfulness, meaning that organizations shouldn't withhold information about why or what they are collecting data. And that they should be clear to the data subjects about their identity and why and how they are processing their data.

- Purpose limitation: It means that data should be collected only for a specific reason. The purpose of processing the data needs to be well-defined to the data subjects.
- Data minimization: Data users should only collect the smallest amount of needed data to complete the processing. There should be a reason to ask about specific data. For example, there is no need to ask for a phone number or address if it's not related to data processing.
- Accuracy: It's up to the organization to deal with data accuracy. Any incorrect or incomplete data stored should be corrected, updated, or erased.
- Storage limitation: There is a need to justify the time the data is being stored. Data retention shouldn't occur if there is no longer a need for the data.
- Integrity and confidentiality: Like the CIA triad, the GDPR requires all data users to maintain the integrity and confidentiality of the collected data. And to ensure its security from unauthorized access.
- Accountability: It's the last principle, which means that the GDPR requires a level of accountability for all organizations. it signifies that documentation and proofs should be associated with the data processing principles. At any time, Supervisory authorities can ask for those documents.

3.4.4.2 2020: The California Consumer Privacy Act (CCPA)

The California Consumer Privacy Act (CCPA) [16] is a state statute intended to regulate how businesses handle the personal information of residents of the state of California. The CCPA was signed into law in 2018 and went into effect on January 1, 2020. As for the CCPA principle, they are as followed:

- Consumers have the right to access all the data an organization collects about them.
- Consumers can choose not to have their information sold to third parties.
- Consumers can demand that organizations delete their personal data.
- Consumers have the right to know to whom their data have been sold to.
- Consumers have the right to know the reason for the data collection.

3.5 IoT SECURITY ISSUES TO PRIVACY

Despite their widespread usage, IoT devices lack security, which is considered the most crucial feature nowadays [9]. Since its origins, security experts have warned about the lack and potential security risk of connected devices to the Internet [67].

Since the late 90s, many attacks have occurred on IoT devices, from hackers infiltrating baby monitors and talking to children to refrigerators and TVs being used to send spam. But the problem is larger than that because the hack of many IoT devices is not to target the devices themselves but to use them as an entry point to the network and the data. From then attack only kept going. A well-known example is the Stuxnet virus [22] which was discovered in 2010. Using malware to infect instructions sent by programmable logic controllers, this virus managed to physically damage Iranian centrifuges. It is reported that this attack started in 2006. The primary occurrence happening in 2009 was one of the first examples of an IoT attack, targeting data acquisition systems and supervisory control in industrial control systems in industrial control systems concerns were raised by implementation were not made. Just three years later, in In December 2013, the first IoT botnet was discovered at enterprise security firm Proofpoint Inc. 25% of it was made of devices rather than a computer such as household appliances, baby monitors, and smart TVs [38].

"I was driving 70 mph on the edge of downtown St. Louis when the exploit began to take hold. Though I hadn't touched the dashboard, the vents in the Jeep Cherokee started blasting cold air at the maximum setting, chilling the sweat on my back through the in-seat climate control system. Next, the radio switched to the local hip-hop station. I spun the control knob left and hit the power button to no avail. Then the windshield wipers turned on, and wiper fluid blurred the glass" [27]. Those were the words of a jeep driver when two security researchers made his part of an experiment. They manage to take complete control of a car remotely. They changed the radio station, turned on the windshield wipers and air conditioner, and even stopped the accelerator from working. It was not made during the experiment due to high risks, but they declared that they could kill the engine and control the brakes. All of this was managed using Uconnect, the in-vehicle connectivity system.

The most significant attack used against IoT devices was in 2016. Where hackers attacked countless websites starting from journalist Brian Krebs' website and OVH web host to Dyn's network, a domain name system caused the unavailability of a website such as Netflix, Amazon, The *New York Times*, and Twitter for hours. This attack was made using the Mirai Botnet by infiltrating consumer IoT devices, IP cameras, and routers [42].

In 2017, the Food and Drug Administration declared that many health-based IoT devices, including pacemakers, cardiac devices, defibrillators, and others, were vulnerable to security intrusions and attacks.

And Lately, in 2021, a group of Swiss hackers was able to hack more than 150,000 live cameras used in monitoring activities inside prisons, schools, hospitals, and private companies such as Tesla.

All of those examples confirm that security is a big issue in IoT systems, hence the user's privacy. Hackers could easily infiltrate those devices, getting access to your data and causing severe damage. As stated in Table 3.3, each of the previous IoT applications has a vulnerability that could impact your privacy.

TABLE 3.3
IoT Application Area, Vulnerabilities, and Impact

	Vulnerabilities	Impact
Smart Home	• Limited AAA • Lack of cryptographic support • Insecure web interface	• Data Loss • Data Corruption • Dos Attack • SQL Injection • Compromise of Personal Information
Smart City	• Limited privacy • Insecure device connectivity • Insecure cloud connectivity	• Brute Force Attack • SSL Man in Middle Attack • SQL injection
Smart Health	• Limited privacy • Insecure device connectivity • Insecure cloud connectivity • Insecure Mobile connectivity • Limited Availability	• Information Disclosure • Dos Attack • Compromise of • Personal Information
Smart Retails	• Insecure device connectivity • Insecure cloud connectivity	• Data Loss • Data Corrumption • Dos Attack • Ransomware
Transportation	• Limited Privacy • Insecure cloud connectivity	• Data Loss • Data Corrumption • Dos Attack

3.5.1 Technologies to Enhance Security Issues

Security should be the number one concern of IoT. Many experts are working on creating a balance between performance and security by using rising new technologies. Below we will discuss how other technologies could enhance the security of IoT and preserve our privacy.

3.5.1.1 Blockchain

Blockchain is a " peer-to-peer" decentralized ledger technology. It provides a method to publicly record and distribute information about transactions on a peer-to-peer system of computers through the crypto protocol. The database is scattered on the rule that each duplicate of new information isn't just put away on a solitary server but shipped off to all clients in the chain or framework. To change any bit of the data set, programmers need to change 51% of the duplicates of the passage on the framework, and every one of these duplicates needs to contain the entire past exchanges of this information. This convention has effectively wiped-out outsiders while guaranteeing security and adaptability with high intuitiveness [51].

Mainly it exists two types of blockchain, public and private. However, there are other variations, too; they consist of consortium and hybrid blockchain. They all consist of clusters of nodes functioning on a peer-to-peer (P2P) network system. Where each node possesses a replica of the shared ledger with timely updates. Also, they all verify, initiate, and recover transactions while also creating new blocks. Blockchain technology has proven itself one of the most secure technologies, especially in the financial world, due to its decentralized nature, transparency, public key-based identity verification, hash-based identity, and consensus-based data creation. All those features enable potential solutions for preserving the data privacy of IoT devices.

Due to the lack of security concerns in IoT, merging it with blockchain technology can reduce those issues and create new security attributes that can enhance the data's privacy and maintain the IoT's proper function.

Below are some of the keys to using Blockchain to secure data privacy:

Private blockchain, as shown in Figure 3.5 for data storing, allows only authorized entities (green nodes) to modify, add, or access the recorded data. In this case, only authorized IoT devices can add data to the blockchain. Which prevents the data from being misused once it has been recorded. Even if an unauthorized party can access the data, the contents will be useless as the data is encrypted with keys. Xu et al. [75] presented a large-scale blockchain-based storage system, called Sapphire for IoT devices that secure the privacy of the data and perform data analytics. Also,

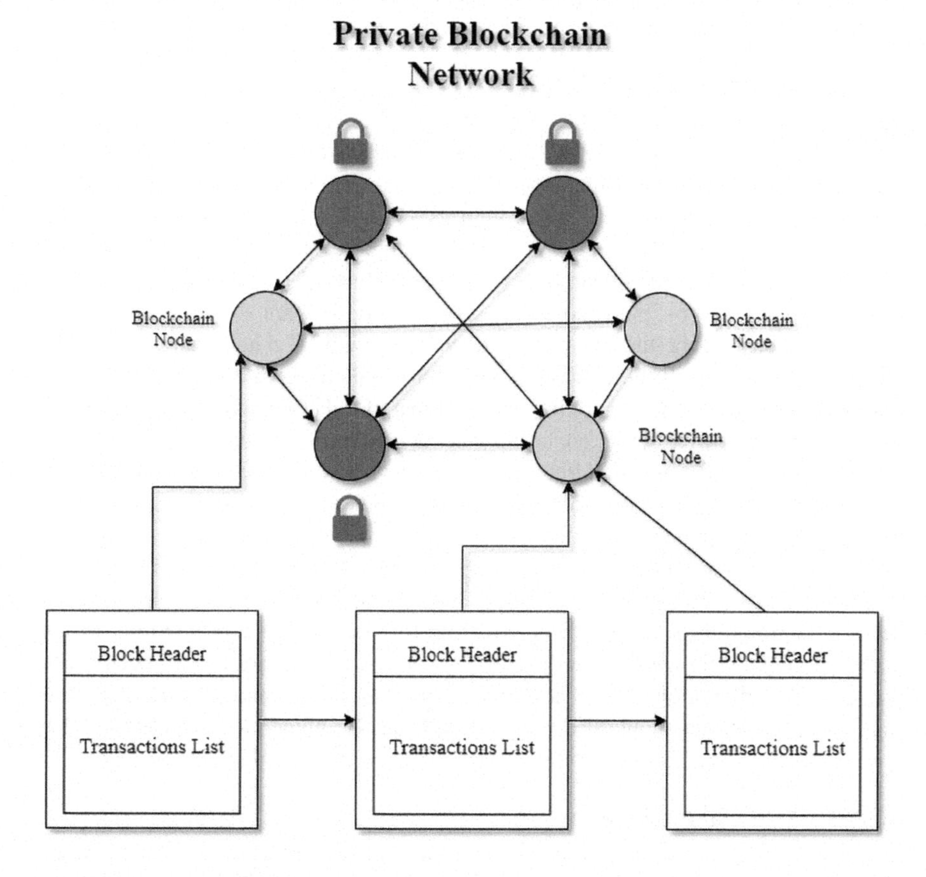

FIGURE 3.5 Private blockchain network.

by storing the data inside a blockchain, we can avoid the problem of single-point failure. Network failure could occur between different devices due to a lack of trust since the IoT ecosystem relies on a centralized server. A study by Lv et al. [47] showed that blockchain could enhance privacy and avoid single-point failure of IoT with a secure, lightweight Publish/Subscribe Model scheme inside a blockchain.

Encryption and miners: We can only store its hash since only 256-bit of the hash key is stored inside the blockchain rather than storing the actual data. And, with the large scale of the data collected by IoT, the storage capacity of the block will not be affected in this case. The actual data could be stored inside an authorized data center where only permission access is given. Hence, the hash will also change if the data is altered, creating secure and private data usage. Moreover, the data stored is verified by miners, which helps avoid corrupted data that could affect the ecosystem.

Data Loss: Once the data enter the blockchain, it cannot be removed or changed. This comes with excellent help for IoT. Especially when IoT devices are facing countless cybersecurity attacks, such as spoofing attacks, where the hacker could access the IoT network data by injecting a node that imitates the network behavior, in this case, it becomes easy to tamper with the data and observe it. Blockchain could prevent such attacks, considering that it can register each authorized user or device, which helps identify malicious devices without the need for central brokers or certification authorities.

3.5.1.2 InterPlanetary File System

Interplanetary File System (IPFS): It's a peer-to-peer (p2p) file-sharing system that aims to change how information is distributed globally. Content is open through peers, found anyplace on the planet that may transfer data, store it, or do both. IPFS realizes how to discover what you request through its substance address instead of its area IPFS [34]. Specifically, IPFS gives a high throughput content that tended to impede the capacity model, with content managed to hyperlinks [53]. Instead of referring to objects (articles, videos, pictures) by the server where they are stored, IPFS refers to them by the hash of the file. By surfing the browser with IPFS and wanting to access a page, IPFS checks all the network nodes even if one of them contains the hash. If a node returns, we get access to the page. Compared to HTTP, where the identifier is the location, finding the server containing the file and retrieving it is easy. In IPFS, we identify the file with content addressing and ask the network where the hash of the file is contained to get it. This is valuable in disconnected cases or in enormous circulated situations where you need to limit load across the organization, giving it an edge over HTTP. Figure 3.6 provides more details explanations of the workflow of IPFS.

All the above features could be used to provide more secure usage of IoT devices, thus, the privacy of the data. IPFS helps store IoT data in an encrypted manner and enable large data storage into small-sized files, where only an authorized person who possesses the encryption key can access the data. Plus, it is decentralized, hence creating P2P verification without third-party interference. Combined with blockchain, they create a decentralized ecosystem for securing data as mentioned by Ali et al. [4].

3.5.1.3 Artificial Intelligence

Artificial intelligence systems can spot patterns and anomalies in ways that humans can't, help cybersecurity deal with cyberattacks that might steal personal data.

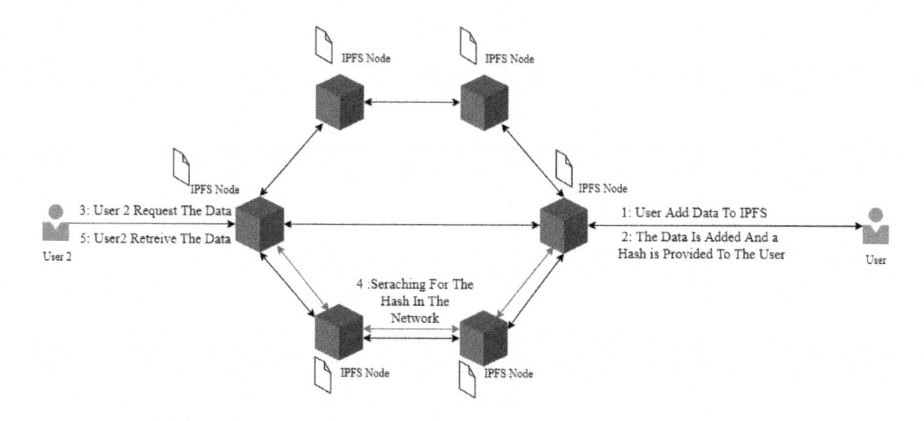

FIGURE 3.6 IPFS workflow.

Machine learning (ML), a technique used in AI, can monitor IoT devices and network activity to detect behavior that's out of the ordinary in real-time and provide protective measures instantly [29]. Models that adapt daily to keep track of new and upcoming threats make them ideal for protecting against complex threats. Only a network using ML systems can instantly detect threats before breaking the network via IoT devices.

Moreover, given many IoT devices, ML could identify external devices that try to retrieve or send data when they are hidden inside the network from a security perspective. They can do that by constantly scanning and comparing historical network behavior. Seeing an unusual increase in network traffic at a particular location, they could provide an alert or act on its [32]. In Table 3.4 we can see different ML models that can be used against IoT cyberattacks.

TABLE 3.4
Example of Machine Learning Models Used against IoT Cyberattacks

Cyberattack	Security Approach	Sutuble ML Models	Frequency
Denial-of-Service	• Access Control • IoT offloading	• Neural Network • Multivariate correlation analysis	High
Intrusion	• Access Control	• Support Vector Machine • KNN • Naive Bayes • Neural Network	Moderate
Spoofing	• Authentication	• Support Vector Machine • Deep Neural Network • Q-learning	High
Malware	• Intrusion Detection System • Access Control	• Random Forest • KNN	Moderate
Jamming	• IoT offloading	• Q-learning	High
Eavesdropping	• Authentication	• Q-learning • Nonparametric Bayesian	High

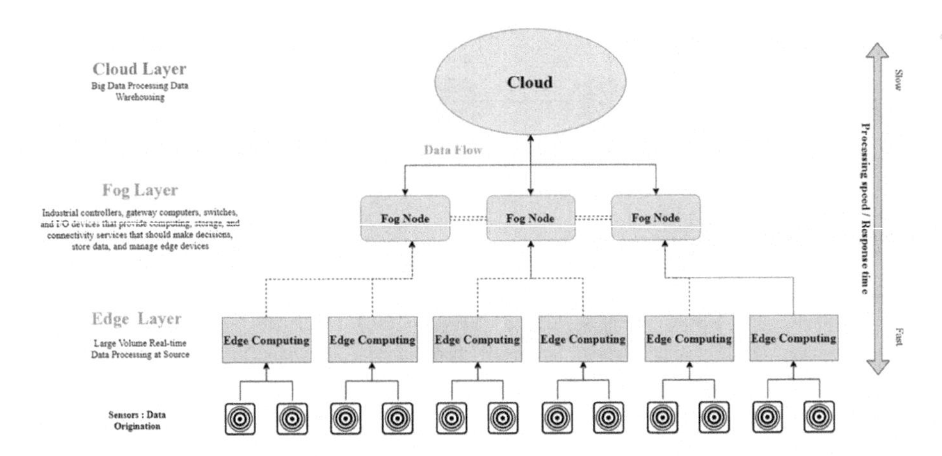

FIGURE 3.7 IoT fog layer architecture.

3.5.1.4 Fog Computing

Fog is another layer of a distributed network environment closely associated with cloud computing. It stands for decentralized computing infrastructure that is located between the cloud and the data source, Figure 3.7 represents the IoT fog computing architecture, which is composed of three layers starting with the edge layer that is directly connected to the sensors, the Fog layer, and finally the cloud layer. It compiles low-latency network connections between devices and processing endpoints. Fog nodes play a significant role by mitigating the impact of low latency and many IoT devices' locations. Do it the relative distance to the data source, storing time-sensitive data becomes more accessible, reducing the impact of privacy issues. Data dissemination is a more critical privacy feature since the data cannot be transferred to the fog node without encryption. Once it is encrypted, searching, or retrieving the cyphertext is very difficult. Moreover, it performs data aggregation to prevent and reduce data leakage and communications [30].

3.6 CORPORATE VIOLATION

There is no denying the benefit that the IoT provides to the human race. As discussed previously, it spreads over every sector, making life more manageable and "fun." But at what cost? Many techniques are being used to adjust the security issues related to IoT devices. However, those concerns only relate to cyberattacks and data breaches caused by cybercriminals, but how about the companies collecting those data? We may say that policies, regulations, and laws are acting to reduce our data usage and that consumers should be offered complete control and transparency over their data. Is it the case? Big companies always find a way because their value depends on your data. They don't expect to make a profit by selling you a wearable smartwatch, a smart home mentor, or any other device that can connect to their server. Their profit is in Big Data. The sensors collect and report everything that is considered part of the Big Data. They are learning and analyzing every single move you make. It's like giving your neighbor every single detail about your life. Yeah, it's enjoyable to go out with him and have fun, but are you willing to provide him with the time you go to bed, the stores you visited during the last few months, the addressee you visited, your medical record, and so on? Personal information should be personal, and that's why IoT is dangerous. It's because you and I are surrendering our privacy, bit by bit, without realizing it. Companies know about us more than we know about ourselves. Buying or using Internet-connected devices is becoming more and more of an obligation. Even if we don't buy, the necessary services require their usage, which could compromise our privacy. Most organizations are trying to comply with these issues by creating their policies. But most people do not read privacy policies for each device they buy or every app they download. Most of it would be written in a legal language incomprehensible to the average consumer, even if they do. They also come with incomprehensible terms of use and clauses that oblige the user to withdraw any right to pursue the company in court if they use their data. For example, in 2013, an Electronic Frontier Foundation activist named Parker Higgins tweeted a phrase he found in the Samsung Smart Tv privacy policy. It read as follows: "Please be aware that if your spoken words include personal or other sensitives information, that information will be among the data captured and transmitted to a third through your use of Voice Recognition "Despite this, you can never be held Samsung accountable since they also state in their policy this sentence." Please note that when you watch a video or access applications or content provided by a third-party, that provider may collect or receive information about your Smart TV (its IP address and device identifiers), the requested transaction (your request to buy or rent the video), and your use of the application or service. Samsung is not responsible for these providers' privacy or security practices. You should exercise caution and review the privacy statements applicable to the third-party websites and services you use." We can only imagine that all of this results in consumer privacy being violated.

Again, IoT is a game-changer in society, and we arrived at a point of no return, meaning that those devices are not going anywhere, and we need to admit that we can not separate ourselves from them. But also, our privacy is equally important, and we should not give up on it quickly. Increased corporate transparency is more than ever, and it's the key to increasing privacy in the IoT. To reach this transparency, people should care more about their privacy and demand more and more details of the usage of data and meaningful consent before collecting them so that industries start by self-regulation and first government to take further actions and severe punishments. Because this is such a complex problem and a fundamental human right is a stack, everyone should be involved. Starting with the consumer, who should demand the right to control his data as he sees fit. Because industries will have no choice but to respond because confidence and trust enable their business to keep growing and existing, it will affect the consumers' desire to purchase connected products, reducing trade and the fulfillment of IoT potential. And most cases, companies tend to respond to those demands, for example, after receiving a survey showing the complaints of newly connected car owners about their concern for their data privacy. The Alliance of Automobile Manufacturers applied and developed new privacy policies to use the collected data. Moreover, in 2020, a consumer survey by Venky Anant [68] was conducted by McKinsey about attitudes toward privacy and data collection. It noted that consumers are becoming more intentional about the type and who they share their data and the connected device they use since no company received more than 50% rate of trust of data protection. However, they indicate that customers are comfortable with companies that provide clear Consent-based data, such as the healthcare and financial industry. This shows that raising the concern of privacy by the customer enhances industry self-regulation, which is beneficial when each company could create specific, standard policies to their client needs and the sensitivity of the data they collect. Also, it makes more interest in Consent-based data. The key is getting the consumer to see the request in their interest.

As for the companies, they should adapt to Layered privacy policies [26]. Those Creative Commons licenses could provide useful models for Consent-based data. Those licenses are constructed of three layers:

- The "legal code" layer: that represents the actual policy, written by lawyers and interpreted by judges.
- The "human-readable" layer should be a simplified summary of the privacy policy in plain language that an average consumer could read.
- The "machine-readable" layer: Is the code that software, search engines, and other kinds of technology can understand and would only allow the technology to have access to information permitted by the consumer.

As those Creative Commons licenses have been used in the past to protect the company's self-interest, they should now deploy them for their customer-interest [25], to create a trust-based interaction, and this could lead to customers being more comfortable with being IoT devices.

Finally, as mentioned previously, governance regulation and laws are essential to reach the required level of privacy that a consumer of a connected device should obtain. With the recent introduction of the EU GDPR and CCPA acting as the first line of defense that conduct specific standards regarding an organization processing "personal data" or " personal information," many data-driven companies are getting more strategic about the data they collect. They can no longer afford to treat each new privacy regulation as a standalone project. Moreover, especially in IoT devices where personal information varies, such as temperature or energy usage in a home, heart rate, body temperature, movement, voice, facial recordings, behavioral patterns, geolocation data, and more, are well noticed in those regulations. We can acknowledge that those regulations contribute to our privacy. True, there are still the same issues they don't fully cover, such as when some companies oblige the user to provide their data without consent or don't get access to their service. But yet, with the rise of new devices and services, those regulations will also adapt and rise [8]. For now, we could be satisfied with what they are offering. As an example, soon after the GDPR took action in 2018, they forced HM to pay €35.3 million to the ICO for illegally surveilling employees at its Nuremberg office in Germany, and in 2020, Marriott International was obliged to pay $23.8 million to the Information Commissioner's Office for compromising its customers' transaction histories and other personal information due to a data breach that occurred in 2018 [60]. As for CCPA, there are no reported fines, but active cases already exist against some of the biggest companies, such as Amazon, Zoom, TikTok, and many others. Despite the significant threat that they impose to companies processing unnecessarily private data without the owner's consent or the ones vulnerable to a data breach, they provide a greater opportunity to improve IoT devices since their distribution is becoming less welcomed by costumer with data privacy at the forefront of their minds, regaining the trust throw those laws will lead to more customer adaptation [21].

3.7 CONCLUSION FUTURE WORK

IoT can assist with making individuals' lives easier, yet we want to ensure it doesn't come at an extreme price to individuals' privacy. But do we believe that they are a necessity for the greater good? "Yes," and do we need to preserve our privacy? "Yes." Therefore privacy balance is struck between people. We consider that having more connected devices enhancing our life is more important than being afraid of losing privacy and those who don't. Despite this, the leading thing is to concede an individual's right to privacy as a fundamental human right. The right to choose how I control my data? How is it being processed? And why are you collecting it? Going through all private data collected by IoT devices may be an enormous undertaking, but it will eventually be worth the effort. Since people are becoming more aware of what they are sharing, what device they are using, and to which company it belongs. True, there is always the risk of cyberattacks, but will it ever end? The rise of new technologies is equal to the height of new threats. Even if we apply the most complex security features, hackers will always find a way to use the same technologies against us. The good thing is that this matter also concerns the organization itself since it deals with significant losses, so we can always count on them trying to fix this issue; why it's a good thing because they are also responsible for privacy violations. With the rise of Big Data, companies are trying to collect as much information as possible to promote their business to a greater dimension. Creating profiles, analyzing consumers' behavior, and tracking every move we make are all part of the big picture, and taking our personal life and transforming it into profit. At least now, the concern is becoming real, and people are beginning to understand the magnitude of the situation and act on it. Governance also starts taking action creating more specific and strict privacy laws. Companies' policies are beginning to change, but it will take more time to reach full consent regarding our data privacy. At least we are on the right path.

In this survey, we summarized the IoT concept from its creation to the near future. We highlighted the major technologies that made IoT what it is today, the evolution of its architecture, and the variety of its usage. But most importantly, we emphasized privacy issues in smart devices, either as security threats or corporate violations. Moreover, we highlighted the solutions that are being implemented to reduce this issue, such as technological solutions to enhance security systems inside IoT devices implementation to create more difficulties for hackers to infiltrate our data and the human behavior and privacy regulations that should cease the collection and processing of personal data by companies when providing a service.

Since the GDPR is strengthening its influence across the globe, future work should focus on its and other rising privacy regulations' impact in the future. In 2021, there was only a 7% increase in companies complaining about GDPR compliance compared to 2020. This means we cannot really analyze its impact on privacy just yet. But it is creating a change, where now we are witnessing the creation of multiple privacy laws in different countries. Like China's Personal Information Protection Law (2020) [54], which is the first initiative made by the Chinese government regarding data privacy. It will take effect by 2021, and it's considered this will increase the legal bases for data processing beyond consent. Furthermore, we have the General Data Protection Law (LGPD) [64] in Brazil which aims to create regulations for Internet users in Latin American countries. It will list ten principles for data privacy control that each sector, private or public, and offline and online organizations should apply. Finally, Canada's Consumer Privacy Protection Act will replace the Personal Information Protection and Electronic Documents Act. It will be a compulsory strict privacy regulation on Canadian organizations. So, in the future, we should focus more on each regulation, on how they are global, affecting data privacy at a national and global level. Also, to study which organization is dealing with those regulations and their adaptation level. what are the new industrial challenges with those regulations? How will it affect the IoT and create new problems?

REFERENCES

[1] A. I. Abdulla, A. S. Abdulraheem, A. A. Salih, M. A. M. Sadeeq, A. J. Ahmed, B. M. Ferzor, O. Salih, and S. I. Mohammed. Internet of Things and Smart Home Security. 62(05):13, 2020.

[2] M. Ahmad, T. Younis, M. A. Habib, R. Ashraf, and S. H. Ahmed. A Review of Current Security Issues in Internet of Things. In M. A. Jan, F. Khan, and M. Alam, editors, *Recent Trends and Advances in Wireless and IoT-Enabled Networks*, EAI/Springer Innovations in Communication and Computing, pages 11–23. Springer International Publishing, Cham, 2019. ISBN 978-3-319-99966-1. https://doi.org/10.1007/978-3-319-99966-1_2.

[3] A. H. Alavi, P. Jiao, W. G. Buttlar, and N. Lajnef. Internet of Things-enabled smart cities: State-of-the-art and future trends. *Measurement*, 129:589–606, December 2018. ISSN 0263-2241. https://www.sciencedirect.com/science/article/pii/S0263224118306912.

[4] M. S. Ali, K. Dolui, and F. Antonelli. IoT Data Privacy via Blockchains and IPFS. In *Proceedings of the Seventh International Conference on the Internet of Things*, IoT '17, pages 1–7, New York, NY, USA, October 2017. Association for Computing Machinery. ISBN 978-1-4503-5318-2. https://doi.org/10.1145/3131542.3131563.

[5] Z. A. Almusaylim and N. Zaman. A Review on Smart Home Present State and Challenges: Linked to Context-Awareness Internet of Things (IoT). *Wireless Networks*, 25(6):3193–3204, August 2019. ISSN 1572-8196. https://doi.org/10.1007/s11276-018-1712-5.

[6] P. Asghari, A. M. Rahmani, and H. H. S. Javadi. Internet of Things Applications: A Systematic Review. *Computer Networks*, 148:241–261, January 2019. ISSN 1389-1286. https://www.sciencedirect.com/science/article/pii/S1389128618305127.

[7] K. Ashton. That 'Internet of Things' Thing. page 1.

[8] C. Barrett. Are the EU GDPR and the California CCPA Becoming the De Facto Global Standards for Data Privacy and Protection? *Scitech Lawyer*, 15(3):24–29, 2019. ISSN 15502090. https://www.proquest.com/docview/2199825726/abstract/2A16470CFBAA43A5PQ/1.

[9] M. Burhan, R. A. Rehman, B. Khan, and B.-S. Kim. IoT Elements, Layered Architectures and Security Issues: A Comprehensive Survey. *Sensors*, 18(9):2796–1424, September 2018. ISSN 1424-8220. https://www.mdpi.com/1424-8220/18/9/2796.

[10] M. Capra, R. Peloso, G. Masera, M. Ruo Roch, and M. Martina. Edge Computing: A Survey on the Hardware Requirements in the Internet of Things World. *Future Internet*, 11(4):100, April 2019. ISSN 1999-5903. https://www.mdpi.com/1999-5903/11/4/100.

[11] F. Caro and R. Sadr. The Internet of Things (IoT) in Retail: Bridging Supply and Demand. *Business Horizons*, 62(1):47–54, January 2019. ISSN 0007-6813. https://www.sciencedirect.com/science/article/pii/S000768131830137X.

[12] S. Chabridon, R. Laborde, T. Desprats, A. Oglaza, P. Marie, and S. M. Marquez. A Survey on Addressing Privacy Together with Quality of Context for Context Management in the Internet of Things. *Annals of Telecommunications – annales des télécommunications*, 69(1):47–62, February 2014. ISSN 1958-9395. https://doi.org/10.1007/s12243-013-0387-2.

[13] L. Chettri and R. Bera. A Comprehensive Survey on Internet of Things (IoT) Toward 5G Wireless Systems. *IEEE Internet of Things Journal*, 7(1):16–32, January 2020. ISSN 2327-4662.

[14] L. Columbus. Roundup of Internet of Things Forecasts and Market Estimates, 2016. https://www.forbes.com/sites/louiscolumbus/2016/11/27/roundup-of-internet-of-things-forecasts-and-market-estimates-2016/?sh=58a0ed39292d. Accessed: 2021-02-10.

[15] D. Cook, M. Youngblood, E. Heierman, K. Gopalratnam, S. Rao, A. Litvin, and F. Khawaja. MavHome: an agent-based smart home. In *Proceedings of the First IEEE International Conference on Pervasive Computing and Communications, 2003. (PerCom 2003)*, pages 521–524, March 2003.

[16] L. de la Torre. A Guide to the California Consumer Privacy Act of 2018. SSRN Scholarly Paper ID 3275571, Social Science Research Network, Rochester, NY, November 2018. https://papers.ssrn.com/abstract=3275571.

[17] M. Degeling, C. Utz, C. Lentzsch, H. Hosseini, F. Schaub, and T. Holz. We Value Your Privacy ... Now Take Some Cookies: Measuring the GDPR's Impact on Web Privacy. *Proceedings 2019 Network and Distributed System Security Symposium*, 2019. http://arxiv.org/abs/1808.05096. arXiv: 1808.05096.

[18] M. M. Dhanvijay and S. C. Patil. Internet of Things: A Survey of Enabling Technologies in Healthcare and Its Applications. *Computer Networks*, 153:113–131, April 2019. ISSN 1389-1286. https://www.sciencedirect.com/science/article/pii/S1389128619302695.

[19] J. Dizdarević, F. Carpio, A. Jukan, and X. Masip-Bruin. A Survey of Communication Protocols for Internet of Things and Related Challenges of Fog and Cloud Computing Integration. *ACM Computing Surveys*, 51(6):116:1–116:29, January 2019. ISSN 0360-0300. https://doi.org/10.1145/3292674.

[20] P. J. Escamilla-Ambrosio, A. Rodríguez-Mota, E. Aguirre-Anaya, R. Acosta- Bermejo, and M. Salinas-Rosales. Distributing Computing in the Internet of Things: Cloud, Fog and Edge Computing Overview. In Y. Maldonado, L. Trujillo, O. Schütze, A. Riccardi, and M. Vasile, editors, *NEO 2016: Results of the Numerical and Evolutionary Optimization Workshop NEO 2016 and the NEO Cities 2016 Workshop held on September 20-24, 2016 in Tlalnepantla, Mexico*, Studies in Computational Intelligence, pages 87–115. Springer International Publishing, Cham, 2018. ISBN 978-3-319-64063-1. https://doi.org/10.1007/978-3-319-64063-1_4.

[21] A. Faison. TikTok Might Stop: Why the IEEPA Cannot Regulate Personal Data Privacy and the Need for a Comprehensive Solution. *Duke Journal of Constitutional Law and Public Policy Sidebar*, 16:115, 2021. https://heinonline.org/HOL/Page?handle=hein.journals/dukjppsid16id=115div=collection

[22] J. P. Farwell and R. Rohozinski. Stuxnet and the Future of Cyber War. *Survival*, 53(1): 23–40, February 2011. ISSN 0039-6338. https://doi.org/10.1080/00396338.2011.555586.

[23] A. Garg, N. Mittal, and Diksha. A Security and Confidentiality Survey in Wireless Internet of Things (IoT). In V. E. Balas, V. K. Solanki, and R. Kumar, editors, *Internet of Things and Big Data Applications: Recent Advances and Challenges*, Intelligent

Systems Reference Library, pages 65–88. Springer International Publishing, Cham, 2020. ISBN 978-3-030-39119-5. https://doi.org/10.1007/978-3-030-39119-5_5.

[24] M. Gascó-Hernandez. Building a Smart City: Lessons from Barcelona. *Communications of the ACM*, 61(4):50–57, March 2018. ISSN 0001-0782. https://doi.org/10.1145/3117800.

[25] A. Gerl. Extending Layered Privacy Language to Support Privacy Icons for a Personal Privacy Policy User Interface. July 2018. https://www.scienceopen.com/hosted-document?doi=10.14236/ewic/HCI2018.177. Publisher: BCS Learning & Development.

[26] A. Gerl and B. Meier. The Layered Privacy Language Art. 12–14 GDPR Extension – Privacy Enhancing User Interfaces. *Datenschutz und Datensicherheit - DuD*, 43(12): 747–752, December 2019. ISSN 1862-2607. https://doi.org/10.1007/s11623-019-1200-9.

[27] A. Greenberg. Hackers Remotely Kill a Jeep on the Highway—With Me in It. https://www.wired.com/2015/07/hackers-remotely-kill-jeep-highway/, 2015. Accessed: 2021-02-09.

[28] S. Hamdan, M. Ayyash, and S. Almajali. Edge-Computing Architectures for Internet of Things Applications: A Survey. *Sensors*, 20(22):6441, January 2020. ISSN 1424-8220. https://www.mdpi.com/1424-8220/20/22/6441.

[29] M. Hasan, M. M. Islam, M. I. I. Zarif, and M. M. A. Hashem. Attack and Anomaly Detection in IoT Sensors in IoT Sites Using Machine Learning Approaches. *Internet of Things*, 7:100059, September 2019. ISSN 2542-6605. https://www.sciencedirect.com/science/article/pii/S2542660519300241.

[30] V. Hassija, V. Chamola, V. Saxena, D. Jain, P. Goyal, and B. Sikdar. A Survey on IoT Security: Application Areas, Security Threats, and Solution Architectures. *IEEE Access*, 7:82721–82743, 2019. ISSN 2169-3536. https://ieeexplore.ieee.org/document/8742551/.

[31] J. Huang, L. Kong, G. Chen, M.-Y. Wu, X. Liu, and P. Zeng. Towards Secure Industrial IoT: Blockchain System with Credit-Based Consensus Mechanism. *IEEE Transactions on Industrial Informatics*, 15(6):3680–3689, June 2019. ISSN 19410050.

[32] F. Hussain, R. Hussain, S. A. Hassan, and E. Hossain. Machine Learning in IoT Security: Current Solutions and Future Challenges. *IEEE Communications Surveys Tutorials*, 22(3):1686–1721, 2020. ISSN 1553-877X.

[33] M. G. Institute. The Internet of Things: Mapping the Value Beyond the Hype. https://www.mckinsey.com/~/media/mckinsey/industries/technology%20media%20and%20telecommunications/high%20tech/our%20insights/the%20internet%20of%20things%20the%20value%20of%20digitizing%20the%20physical%20world/the-internet-of-things-mapping-the-value-beyond-the-hype.ashx

[34] IPFS. IPFS Documentation. https://docs.ipfs.tech/. Accessed: 2023-05-10.

[35] J. Isaak and M. J. Hanna. User Data Privacy: Facebook, Cambridge Analytica, and Privacy Protection. *Computer*, 51(8):56–59, August 2018. ISSN 1558-0814.

[36] M. T. Islam, T. Alam, I. Yahya, and M. Cho. Flexible Radio-Frequency Identification (RFID) Tag Antenna for Sensor Applications. *Sensors*, 18(12):4212, December 2018. ISSN 1424-8220. https://www.mdpi.com/1424-8220/18/12/4212.

[37] B. Jan, H. Farman, M. Khan, M. Talha, and I. U. Din. Designing a Smart Transportation System: An Internet of Things and Big Data Approach. *IEEE Wireless Communications*, 26(4):73–79, August 2019. ISSN 1558-0687.

[38] M. A. Jan, P. Nanda, X. He, Z. Tan, and R. P. Liu. A Robust Authentication Scheme for Observing Resources in the Internet of Things Environment. In *2014 IEEE 13th International Conference on Trust, Security and Privacy in Computing and Communications*, pages 205–211, September 2014. ISSN: 2324-9013.

[39] X. Jia, Q. Feng, T. Fan, and Q. Lei. RFID Technology and Its Applications in Internet of Things (IoT). In *2012 2nd International Conference on Consumer Electronics, Communications and Networks (CECNet)*, pages 1282–1285, April 2012.

[40] J. T. Kelly, K. L. Campbell, E. Gong, and P. Scuffham. The Internet of Things: Impact and Implications for Health Care Delivery. *Journal of Medical Internet Research*, 22(11):e20135, November 2020. https://www.jmir.org/2020/11/e20135.

[41] A. Khanna and S. Kaur. Internet of Things (IoT), Applications and Challenges: A Comprehensive Review. *Wireless Personal Communications*, 114(2): 1687–1762, September 2020. ISSN 1572-834X. https://doi.org/10.1007/s11277-020-07446-4.

[42] C. Kolias, G. Kambourakis, A. Stavrou, and J. Voas. DDoS in the IoT: Mirai and Other Botnets. *Computer*, 50(7):80–84, 2017. ISSN 1558-0814.

[43] M. Kuzlu, C. Fair, and O. Guler. Role of Artificial Intelligence in the Internet of Things (IoT) Cybersecurity. *Discover Internet of Things*, 1(1):7, February 2021. ISSN 2730-7239. https://doi.org/10.1007/s43926-020-00001-4.

[44] H. Landaluce, L. Arjona, A. Perallos, F. Falcone, I. Angulo, and F. Muralter. A Review of IoT Sensing Applications and Challenges Using RFID and Wireless Sensor Networks. *Sensors*, 20(9):2495–1424, January 2020. ISSN 1424-8220. https://www.mdpi.com/1424-8220/20/9/2495.

[45] I. Lee. The Internet of Things for Enterprises: An Ecosystem, Architecture, and IoT Service Business Model. *Internet of Things*, 7:100078, September 2019. ISSN 2542-6605. https://www.sciencedirect.com/science/article/pii/S2542660519301386.

[46] S. Li, L. D. Xu, and S. Zhao. 5G Internet of Things: A Survey. *Journal of Industrial Information Integration*, 10:1–9, June 2018. ISSN 2452-414X. https://www.sciencedirect.com/science/article/pii/S2452414X18300037.

[47] P. Lv, L. Wang, H. Zhu, W. Deng, and L. Gu. An IOT Oriented Privacy- Preserving Publish/Subscribe Model Over Blockchains. *IEEE Access*, 7:41309–41314, 2019. ISSN 2169-3536.

[48] M. S. Mahdavinejad, M. Rezvan, M. Barekatain, P. Adibi, P. Barnaghi, and A. P. Sheth. Machine Learning for Internet of Things Data Analysis: A Survey. *Digital Communications and Networks*, 4(3):161–175, August 2018. ISSN 2352-8648. https://www.sciencedirect.com/science/article/pii/S235286481730247X.

[49] J. Mesquita, D. Guimarães, C. Pereira, F. Santos, and L. Almeida. Assessing the ESP8266 WiFi Module for the Internet of Things. In *2018 IEEE 23rd International Conference on Emerging Technologies and Factory Automation (ETFA)*, volume 1, pages 784–791, September 2018. ISSN: 1946-0759.

[50] E. Moyakine. Online Anonymity in the Modern Digital Age: Quest for a Legal Right. *Journal of Information Rights, Policy and Practice*, 1(1), October 2016. https://doi.org/10.21039/irpandp.v1i1.21

[51] Q. K. Nguyen and Q. V. Dang. Blockchain Technology for the Advancement of the Future. In *2018 4th International Conference on Green Technology and Sustainable Development (GTSD)*, pages 483–486, Ho Chi Minh City, November 2018. IEEE. ISBN 978-1-5386-5126-1. https://ieeexplore.ieee.org/document/8595577/.

[52] M. M. H. Onik, C.-S. Kim, and J. Yang. Personal Data Privacy Challenges of the Fourth Industrial Revolution. In *2019 21st International Conference on Advanced Communication Technology (ICACT)*, pages 635–638, February 2019. ISSN: 1738-9445.

[53] M. Pechenizkiy and M. Wojciechowski. *New Trends in Databases and Information Systems*, volume 185 of *Advances in Intelligent Systems and Computing*. Springer Berlin Heidelberg, Berlin, Heidelberg, 2013. ISBN 978-3-642-32517-5 978-3-64232518-2. http://link.springer.com/10.1007/978-3-642-32518-2.

[54] E. Pernot-Leplay. China's Approach on Data Privacy Law: A Third Way between the U.S. and the EU? *Penn State Journal of Law and International Affairs*, 8:49, 2020. https://heinonline.org/HOL/Page?handle=hein.journals/pensalfaw8id=53div=collection=.

[55] Y. A. Qadri, A. Nauman, Y. B. Zikria, A. V. Vasilakos, and S. W. Kim. The Future of Healthcare Internet of Things: A Survey of Emerging Technologies. *IEEE Communications Surveys Tutorials*, 22(2):1121–1167, 2020. ISSN 1553877X.

[56] Y. Qian, D. Wu, W. Bao, and P. Lorenz. The Internet of Things for Smart Cities: Technologies and Applications. *IEEE Network*, 33(2):4–5, March 2019. ISSN 1558156X.

[57] J. Romkey. Toast of the IoT: The 1990 Interop Internet Toaster. *IEEE Consumer Electronics Magazine*, 6(1):116–119, January 2017. ISSN 2162-2256.

[58] D. Sachan, M. Goswami, and P. K. Misra. Analysis of Modulation Schemes for Bluetooth-LE Module for Internet-of-Things (IoT) Applications. In *2018 IEEE International Conference on Consumer Electronics (ICCE)*, pages 1–4, January 2018. ISSN: 2158-4001.

[59] P. Saffo. Sensors: The Next Wave of Innovation. *Communications of the ACM*, 40(2):92–97, February 1997. ISSN 0001-0782, 1557-7317. https://dl.acm.org/doi/10.1145/253671.253734.

[60] B. Sayar. The Administrative Fines Regime of the General Data Protection Regulation and Its Impact. *Kişisel Verileri Koruma Dergisi*, 3(1):1–16, June 2021. ISSN 2667-6524. https://dergipark.org.tr/en/pub/kvkd/issue/62960/852700.

[61] M. Serror, S. Hack, M. Henze, M. Schuba, and K. Wehrle. Challenges and Opportunities in Securing the Industrial Internet of Things. *IEEE Transactions on Industrial Informatics*, 17(5):2985–2996, May 2021. ISSN 1941-0050.

[62] G. Shan and B.-H. Roh. Performance Model for Advanced Neighbor Discovery Process in Bluetooth Low Energy 5.0-Enabled Internet of Things Networks. *IEEE Transactions on Industrial Electronics*, 67(12):10965–10974, December 2020. ISSN 1557-9948.

[63] N. Sharma, M. Shamkuwar, and I. Singh. The History, Present and Future with IoT. In V. E. Balas, V. K. Solanki, R. Kumar, and M. Khari, editors, *Internet of Things and Big Data Analytics for Smart Generation*, Intelligent Systems Reference Library, pages 27–51. Springer International Publishing, Cham, 2019. ISBN 978-3-030-04203-5. https://doi.org/10.1007/978-3-030-04203-5_3.

[64] T. L. Sombra. The General Data Protection Law in Brazil: What Comes Next? *Global Privacy Law Review*, 1(2), June 2020. https://kluwerlawonline.com/journalarticle/Global+Privacy+Law+Review/1.2/GPLR2020083

[65] Symantic. Executive Summary: 2018 Internet Security Threat Report. https://docs.broadcom.com/doc/istr-23-executive-summary-en, 2015.

[66] M. Sysoev, A. Kos, and M. Pogačnik. Noninvasive Stress Recognition Considering the Current Activity. *Personal and Ubiquitous Computing*, 19(7):1045–1052, October 2015. ISSN 1617-4917. https://doi.org/10.1007/s00779-015-0885-5.

[67] A. Tewari and B. B. Gupta. Security, Privacy and Trust of Different Layers in Internet-of-Things (IoTs) Framework. *Future Generation Computer Systems*, 108:909–920, July 2020. ISSN 0167-739X. https://www.sciencedirect.com/science/article/pii/S0167739X17321003.

[68] J. K. H. S. Venky Anant, Lisa Donchak. The Consumer-Data Opportunity and the Privacy Imperative. https://www.mckinsey.com/capabilities/risk-and-resilience/our-insights/the-consumer-data-opportunity-and-the-privacy-imperative. Accessed: 2021-02-09.

[69] P. Voigt and A. von dem Bussche. Enforcement and Fines under the GDPR. In P. Voigt and A. von dem Bussche, editors, *The EU General Data Protection Regulation (GDPR): A Practical Guide*, pages 201–217. Springer International Publishing, Cham, 2017. ISBN 978-3-319-57959-7. https://doi.org/10.1007/978-3-319-57959-7.

[70] P. Voigt and A. von dem Bussche. Introduction and 'Checklist'. In P. Voigt and A. von dem Bussche, editors, *The EU General Data Protection Regulation (GDPR): A Practical Guide*, pages 1–7. Springer International Publishing, Cham, 2017. ISBN 978-3-319-57959-7. https://doi.org/10.1007/978-3-319-57959-7_1.

[71] P. Voigt and A. von dem Bussche. Practical Implementation of the Requirements under the GDPR. In P. Voigt and A. von dem Bussche, editors, *The EU General Data Protection Regulation (GDPR): A Practical Guide*, pages 245–249. Springer International Publishing, Cham, 2017. ISBN 978-3-319-57959-7. https://doi.org/10.1007/978-3-319-57959-7_10.

[72] S. Vučenović. Internet of Things as Innovative Technology in Retailing. *Anali Ekonomskog fakulteta u Subotici*, (39):249–256, 2018. ISSN 0350-2120. http://scindeks.ceon.rs/article.aspx?artid=0350-21201839249V.

[73] R. H. Weber. Internet of Things: Privacy Issues Revisited. *Computer Law & Security Review*, 31(5):618–627, October 2015. ISSN 02673649. https://linkinghub.elsevier.com/retrieve/pii/S0267364915001156.

[74] F. Xia, L. T. Yang, L. Wang, and A. Vinel. Internet of Things. *International Journal of Communication Systems*, 25(9):1101–1102, September 2012. ISSN 10745351. https://onlinelibrary.wiley.com/doi/10.1002/dac.2417.

[75] Q. Xu, K. M. M. Aung, Y. Zhu, and K. L. Yong. A Blockchain-Based Storage System for Data Analytics in the Internet of Things. In R. R. Yager and J. Pascual Espada, editors, *New Advances in the Internet of Things*, volume 715, pages 119–138. Springer International Publishing, Cham, 2018. ISBN 978-3-319-58189-7 978-3-319-58190-3. https://doi.org/10.1007/978-3-319-58190-3_8.

4 Security, Privacy, and Blockchain in Financial Technology[1]

Schallum Pierre
Institut intelligence et données (IID),
Université Laval, Québec, Canada

Olson Italis
Polytechnique Montréal/ISTEAH, Québec, Canada

CONTENTS

4.1 Introduction 72
4.2 Six Technologies for Mobile Payment Systems 73
4.2.1 Short Message Service (SMS) 73
4.2.2 Unstructured Supplementary Service Data (USSD) 73
4.2.3 Bluetooth Low Energy (BLE) or Bluetooth Smart 74
4.2.4 Wireless Application Protocol (WAP) 74
4.2.5 Quick Response Code (QRC) 74
4.2.6 Near Field Communication (NFC) 75
4.3 Security 75
4.3.1 Cyber-attacks and HDM 76
4.3.1.1 Malware and Malicious Domain 77
4.3.1.2 SSL/TLS and HDM Protocols 77
4.3.2 Contactless Payment and Limits (NFC: SE and HCE) 77
4.3.2.1 Phone Data Attack 78
4.3.2.2 Attack between the Application Processor and the NFC Controller on Mobile 78
4.3.3 NFC and the Limit of the Preregistered Card Requirement 79
4.3.4 Relay Attacks 79
4.3.5 Cryptographic Primitives 79
4.3.6 Cloud Computing: Cloudlet for Resource Exchange between Users without Wi-Fi and Limitations 80
4.4 The Privacy Problem 80
4.4.1 Privacy with the Considered Technologies 80
4.4.2 Privacy with Biometric Authentication 82
4.4.2.1 Protection of Stored Biometric Data 83
4.4.2.2 Protection during Exchanges with the Remote Server 83

DOI: 10.1201/9781003227656-7

4.4.2.3 Biometric Data Protection Solutions ... 83
4.4.2.4 Balance between Accuracy and Privacy ... 83
4.5 Protecting Privacy and Enhancing Digital Trust ... 84
4.5.1 Secure Computer System and TrustZone ... 84
4.5.2 Transit Payment System and Privacy Protection ... 84
4.5.3 Asymmetric Encryption and Encrypted QR Code ... 85
4.5.4 Blockchain and Privacy ... 86
4.5.5 Contactless Payment and Privacy ... 87
4.6 Ethical Recommendations for Protecting the Life Cycle of Personal Payment Data ... 88
4.6.1 Collection ... 88
4.6.2 Storage ... 88
4.6.3 Exploitation ... 89
4.6.4 Destruction or Anonymization ... 89
4.7 Conclusion ... 89
References ... 90

4.1 INTRODUCTION

COVID-19 and the multiple variants associated with it have contributed to an acceleration of the digital transformation. One of the most visible areas of this transformation is mobile payments, which are related to financial technologies or fintech. According to Amit Samsukha [1], in 2020, mobile payments have accounted for 44.5% of all e-commerce transactions, double the amount of debit card payments and triple the amount of credit card payments. From 3.53 trillion dollars (USD), in 2018, with its development, it should reach, according to *Fortune Business Insights* [2], 19.89 trillion dollars in 2026. In Québec, for example, the use of in-store mobile payment by adults jumped by 8 percentage points [3], from 17% in 2019, it rose to 25%, in 2021.

During the pandemic, contactless digital payments [4], such as cards or e-wallets, were recommended. The need for remote payments is driving the creation of either new services for some companies or several new players in the financial technology sector. This requires the choice of secure infrastructures to collect, store, and use personal data.

Indeed, data collection represents a significant risk to privacy, especially when it is used to profile citizens. As shown by the European Data Protection Supervisor (EDPS), it can also increase the risk of cyber-attacks [5]. During this pandemic context, the cybersecurity market, including online crime, online threats, and data breaches, exploits to its advantage the vulnerability of individuals and the fear that takes hold in cities [6]. Unprecedented strategies have been used to steal funds and collect private data [7]. In the United States, between February and April 2020, cyber-attacks targeting the financial sector increased by 238% [6]. In the face of this, global spending on cybersecurity will need to increase from $173 billion in 2020 to $270 billion in 2026, according to *Forbes* [8]. This chapter focuses on the problem of the security of personal data generated in digital environments, highlighting the ethical issues related to technologies used in fintech including mobile payment systems. Ethics aims at protecting people. It anticipates potential privacy issues to better prevent them. In the field of mobile payment, ethics considers technological and legal advances. For example, on a global scale, the Payment Card Industry (PCI) Security Standards Council sets rules for the protection of account data [9]. Countries such as Canada have regulations in place to protect personal data including financial data: The Personal Information Protection and Electronic Documents Act (PIPEDA) [10] and the Canadian Payments Act (R.S.C. (1985), c. C-21) [11]. Very recently, Québec adopted Bill 64 (current law 25) to protect personal information [12] and Bill 6, which created the new Ministry of Cybersecurity and Digital Affairs [13]. In the European context, since 2018, member state of the European Union must comply with the General Data Protection Regulation (GDPR) [14]. In the specific context of France, it is framed by the French National Agency for the Security of Information Systems (In French: Agence nationale de la sécurité des systèmes d'information or ANSSI) [15] and the National Commission on Informatics and Liberty (In French: Commission Nationale de l'Informatique et des Libertés or CNIL) [16]. This chapter provides a summary of what is known about the technologies used in mobile payment systems. Given the prominent role that personal data plays in mobile payments, on what basis should a technological choice be made? How can digital trust be improved based on the technologies used in mobile payment systems?

This chapter considers the security issues of mobile payment in its various forms [17] such as the point of sale (POS) system, in-store and remote payments.

It is divided into five sections that list the different technological approaches to mobile payment deployment and their impact on data security. In the first section, six mobile payment system technologies are discussed. In the second section, security issues are considered from the perspective of cyber-attacks. In the third, privacy issues are described. In the fourth, blockchain is examined with respect to privacy. Finally, in the fifth section, recommendations are made for protecting the personal payment data life cycle.

4.2 SIX TECHNOLOGIES FOR MOBILE PAYMENT SYSTEMS

In his book Mobile Payment, Thomas Lerner [18] describes the main technologies used in mobile payment systems, which refers to transactions made via a mobile device [19]. Each of the technologies discussed has advantages and disadvantages [20].

4.2.1 Short Message Service (SMS)

The "Short Message Service" (SMS) is a messaging service allowing users to use up to 160 characters. It was developed for use in the Global System for Mobile Communications (GSM) network. Very simple to use and known throughout the world, it is however expensive and can easily become the target of cyber-attacks because of faulty cryptography techniques.

4.2.2 Unstructured Supplementary Service Data (USSD)

Unstructured Supplementary Service Data (USSD) generally uses the Global System for Mobile Communication/Short Message Service (GSM/SMS) and serves as an interface for the customers themselves and between them and the banks. Its advantage is that it is easy to use and compatible with all mobiles. The disadvantage is that the data is not secure enough. Indeed, the systems usually use a personal identification number (PIN) to authenticate the user at the application level [18], but as in the previous case, the cryptographic procedures are flawed.

4.2.3 Bluetooth Low Energy (BLE) or Bluetooth Smart

Bluetooth Low Energy (BLE) or Bluetooth Smart uses wireless transmission. Because of its low-energy requirements, it has a wide application in the proximity payment sector, hence its limitations as well. It has the advantage of allowing secure exchanges by respecting the recommendations of the National Institute Standard and Technology (NIST). This organization has described two modes and security levels for a service between two devices connected via BLE [21, 22]:

- For mode 1, level 1 does not initiate encryption or authentication, level 2 is only about encryption but not authentication, Level 3 requires authentication and encryption, and level 4 requires the use of AES-CMAC elliptic curve keys of 250 bits in length.
- For mode 2, data signing is considered at two levels depending on whether authentication is required at level 1 or level 2 at the beginning of the connection establishment.

For a service such as mobile payment, where security is required, NIST recommends mode 1 and level 4 where both devices must authenticate and the exchanges encrypted using AES-CMAC with p-256 ECC.

Moreover, BLE is compatible with most smartphones. This is an advantage since a mobile payment system with BLE has high availability.

4.2.4 Wireless Application Protocol (WAP)

The Wireless Application Protocol (WAP) allows a mobile device to access the Internet. Thanks to the integration of the Wireless Transport Layer Security (WTLS), which encrypts the exchanges, the security of the protection of the user's data and the authentication of the server is very high. WAP is especially recommended for online payment. However, its use is still limited, even though there is great interest in it. Furthermore, the technology is outdated and is not suitable for delivering a consistent choice of services to current terminals, which Samuel Pierre had foreseen long ago [23].

4.2.5 Quick Response Code (QRC)

The "Quick Response Code" (QRC) or "black and white matrix barcodes" or "Data Matrix barcode" is a two-dimensional barcode, also known as a 2D code or matrix code. It makes it possible to read numeric, alphanumeric, or binary data with a smartphone equipped with a camera. The Secure QRC (SQRC) has been added to the set-in order to strengthen the encryption measures because it has serious security problems, in particular, the redirection of the user to a malicious site, which is a fake site capable of stealing financial data [24]. It is comparable to BLE in terms of accessibility, as it is available on mid-range phones.

4.2.6 Near Field Communication (NFC)

Near Field Communication (NFC) is used for contactless payment. This radio communication technology uses the 13.56 MHz frequency in the free "Industrial, Scientific and Medical" (ISM) band. It allows data exchange and payment between two devices located at a distance of 4–10 centimeters. It is used in two configurations: one where a "secure element" (SE) is inside the phone – notably within a Subscriber Identity Module (SIM) card – and the other where the SE is integrated into a cloud server called "Host Card Emulation" (HCE). While contactless payment is fast and has been very successful, there are two major drawbacks to consider: the HCE configuration is vulnerable to a relay attack, and with the embedded SE, the service provider is very dependent on the hardware provider and deployment can also be expensive. In the case of the replay attack, an attacker can capture a user's banking information on the fly and send it to another person who can use it to make purchases [25].

The six technologies described above are used in several mobile payment systems, which are discussed in the Sections 4.3 and 4.4. A summary of the respective advantages and limitations of each of these technologies is given in Table 4.1.

The various security issues bring ethics back to the center of mobile payment systems. They concern cyber-attacks, the so-called "Man In The Middle" (MITM) attack, and the relay attack. They are explained in the "security" section.

4.3 SECURITY

Payment transactions generate a considerable amount of data. The white paper published by the CNIL "When trust pays off: today's and tomorrow's means of payment methods facing the challenge of data protection" [26] describes the magnitude of the variety of information that comes from payment data. According to the CNIL, these are:

- Actual payment data
  - date and time of payment
  - identity of the merchant
  - identity of the beneficiary
  - international bank account number or IBAN
- Purchase or checkout data
  - date and place of purchase
  - loyalty card details
- Contextual or behavioral data
  - customer knowledge data
  - geolocation
  - characteristics of the terminal used for an online purchase
  - characteristics of the products explored prior to the purchase
  - the time spent browsing

TABLE 4.1
Technologies in Mobile Payment: Advantages and Disadvantages

Technologies	Advantages	Disadvantages
SMS	• High accessibility: available on GSM networks and with low-end phones.	• Service can be expensive; • Obvious security flaws.
USSD	• Available on GSM networks and with low-end phones; • No cost associated with this service for a mobile network operator.	• Security flaws: cryptographic properties defeated, low-entropy password (4 or 6 digits).
BLE	• Adequate security level for mobile payment; • Medium accessibility: available on mid-range phones; • Ease of deployment of the service.	• Proximity payment only: another communication technology is needed to interconnect remote nodes.
WAP	• Technology that allows mobiles (without adequate capacity) to connect to the Internet; • Use of cryptographic properties.	• Technology outdated by current service offering: mobiles connect to the Internet without the need for another layer.
QRC	• Average accessibility as for BLE; • Ease of deployment comparable to the effort required to deploy a system with BLE.	• Proximity payment only: other communication technologies needed to interconnect remote nodes; • Redirection to malicious sites, a threat to such systems.
NFC	• Architecture with integrated OS very secure; • Great ease of use: contactless payment.	• Expensive service deployment: payment service provider highly dependent on the provider of the embedded security element, the basis of the system's security robustness; • HCE architecture, system vulnerable to relay attacks.

These different forms of data can contribute to the physical identification of a person. Consequently, these data are personal and sensitive. As such, it must be protected as soon as it is generated. Minimal data collection is required [27].

The security of the technology of any mobile payment system – mobile payment at a POS, mobile payment as a POS, mobile payment platform, independent mobile payment system, and direct billing by the operator – is essential. User data is at the heart of the digital economy. It is also subject to cyber-attacks.

4.3.1 Cyber-Attacks and HDM

Mobile payment systems are exposed to different types of threats and cyber-attacks. These include malicious programs or malware and malicious domain.

4.3.1.1 Malware and Malicious Domain

The most important threat, from a security perspective, is malware. Some useful applications – such as those for call recording, instant messaging, Global Positioning System (GPS) tracking, and call log transfer – generate malware on a cell phone. An effective malware detection method dedicated to cell phones will have to be developed. Malware detection methods, attached to static analysis, dynamic analysis, and legal framework for the mobile environment, are currently not effective for mobile devices [28]. In addition to malicious programs or software, during COVID-19, malicious domains or fake websites appeared. Toward the end of March 2020, Palo Alto Networks identified 40,261 high-risk malicious domains and 2,022 new malicious registered domains using keywords such as "covid-19," "covid19," "coronavirus," and "corona-virus" [29]. According to Europol, online criminals are exploiting increasing anxiety, high demand for protective equipment and pharmaceuticals, decreasing mobility, and increasing teleworking to conduct unsolicited communication campaigns (spam) and obtain sensitive information (credit card numbers, login credentials) about organizations (hospitals and international health agencies) and individuals [30]. Preventive measures such as identity protection are necessary to protect personal data from the threat of site hijacking. User data can be hacked by the MITM attack facilitated by the Secure Sockets Layer/Transport Layer Security (SSL/TLS) protocols.

4.3.1.2 SSL/TLS and HDM Protocols

Even though it prevents hackers from accessing sensitive data in transit thanks to the encryption algorithm [31], the SSL protocol (a more secure version of which is TLS) on which many mobile devices rely for their security may have potentially exploitable vulnerabilities. This is the case with the Heartbleed Bug vulnerability that was found in the OpenSSL cryptographic library. It can be exploited by malicious user agents to steal information belonging to the owner of the mobile. The SSL protocol (or TLS for the new version) is also vulnerable to a MITM attack. HDM refers to any attack carried out with the objective of accessing exchanges between two interlocutors on the Internet [32]. The attack can be the source of fraudulent transactions leading to embezzlement. In order to increase the digital trust of the users, it is crucial to implement a security system for the protection of data on the backend through the valid certificate of the server. It will be necessary to develop an effective security system that will automatically terminate any transaction with an invalid certificate [28].

4.3.2 Contactless Payment and Limits (NFC: SE and HCE)

Another technology is contactless payment, which has recently been deployed, on a very large scale, in many banking institutions and retail stores, despite some limitations [33]. In the COVID-19 era, for health reasons, cash exchanges tend to be supplanted by contactless payment [34]. NFC is the technology that enables this contactless payment.

The article "Fraud on host card emulation architecture: Is it possible to fraud a payment transaction realized by a cell phone using an 'Host Card Emulation' system of security?" [35] discusses two main architectures that secure this type of payment: the SE that is integrated into the cell phone and the HCE technology, which is an architecture that is not integrated into the cell phone but connected via a network. The SE is only in the SIM card. This means that the mobile's operating system has no access to the transaction data. This is its strength. However, since the SE is attached to a mobile operator, there is the problem of interoperability. HCE technology facilitates payment via NFC technology, which is found on "more than two out of three smartphones" [36]. HCE is proposed by the electronic money industry to solve the problem of EM interoperability. It proceeds to desynchronize the SE and makes it possible to store banking data on cloud computing. It uses tokenization or the replacement of banking data with a token (an element that represents sensitive data only within the system). The goal is to prevent any form of reuse of sensitive data. However, two types of cyber-attacks have been reported. Firstly, the attack on phone data and secondly the attack between the application processor and the NFC controller on the cell phone [35]. Before choosing SE and HCE, these two important security issues will need to be addressed.

4.3.2.1 Phone Data Attack

With the HCE server, sensitive data is traced back to the mobile's operating system. In order to prevent another cell phone from taking control of the cell phone to which the HCE server is linked via a malicious program, the phone is recognized by its IP address and Media Access Control (MAC) address. The IP address (for Internet Protocol) is a string of characters that uniquely identifies an entity connected to the Internet. This string has a length that depends on the protocol version, IPv4 (version 4) or IPv6 (version 6); the MAC address or physical address is a string of characters that identifies a network interface, it is the address of a subnet in a larger network [37]. When these addresses correspond to those known to the HCE server, the latter proceeds to send the data. There are three elements that can be used to attack a cell phone using HCE technology: the symmetric key, the MAC address, and the IP address. Strengthening these three elements is essential to improve the security of the HCE server [35]. Another vulnerability in contactless payment is the attack between the application processor and the NFC controller.

4.3.2.2 Attack between the Application Processor and the NFC Controller on Mobile

This type of attack occurs in disconnected mode, a mode that allows the payment to be used permanently via HCE. The authorization is possible because the card numbers have been uploaded to the network. Even if the cell phone is disconnected from the network, the payment can be authorized. If a malicious program manages to transfer the important data to another phone, the transaction can be completed without the authentic holder or the HCE server being aware of it. The HCE technology will benefit from being more robust, to avoid malware attacks [25, 35]. Contactless payment faces other important technical and functional constraints.

4.3.3 NFC and the Limit of the Preregistered Card Requirement

In "A prototype-based case study of secure mobile payments" [38], Till Halbach explores the relationship between user interactions, security, and privacy mechanism for a mobile payment case based on NFC technology. The resulting insights can be used to develop payment solutions for Apple and Google systems, for example. An Android application has been developed as a proof of concept. However, there are many technical and functional constraints to making a payment, as a physical card must first be registered before using the application. The bank's requirements must be taken into account. The user cannot choose a particular card during the transaction [38]. Solutions must be found to avoid the limitations of contactless payment, which is also vulnerable to relay attacks.

4.3.4 Relay Attacks

The infrastructure of the open mobile payment platform "SIMulations des Mobilités" or MobiSIM [39] – modeling daily and residential simulations for sustainable planning of French and European cities – is based on HCE technology. It uses Europay Mastercard Visa (EMV) SE cards hosted in cloud computing and remotely accessible via the Remote APDU Call Secure (RACS) protocol. It is described by an Internet Engineering Task Force (IETF) project whose security is based on the Transport Layer Security (TLS) protocol that applies strong mutual authentication and runs in the SIM module. Transactions are completed in less than a second. However, MobiSIM faces relay attacks. In addition to this relay attack problem, the problem of limiting the use of contactless bank cards via the network must be solved. Indeed, these peripheral cards can only perform a limited number of contactless transactions. They require the entry of a PIN after a few dozen payments [40]. As the cell phone is now a tool through which users store several categories of data, it is important to make use of strong communication security as allowed by cryptographic primitives.

4.3.5 Cryptographic Primitives

The article "A Secure Transaction Scheme With Certificateless Cryptographic Primitives for IoT-Based Mobile Payments" [41] introduces a secure transaction scheme with "cryptographic primitives" for mobile payments. The latter are modules that provide cryptographic hash and encryption functions to guarantee the security of computer systems. The proposed scheme takes advantage of the merits of Android Pay and a certified crypto-signature system to simultaneously ensure transaction security and achieve payment efficiency in practice. It is both accurate and secure via a random oracle model. It offers robustness and communication security for mobile users during online payment transactions. On the other hand, the performance evaluation shows the practicality of the proposed transaction scheme as the total computational cost is acceptable for an Internet of Things (IoT)-based test. Due to the adoption of the bilinear matching crypto-operation, the computational performance and scalability of the system is therefore limited. It will be necessary to improve the security components adopted in the proposed scheme, before the choice of this technology. Along with security, there is data management in cloud computing referring to any access to a computing service via the Internet. The cloudlet model proposes a multilateral resource exchange framework for consideration.

4.3.6 Cloud Computing: Cloudlet for Resource Exchange between Users without Wi-Fi and Limitations

Wu and Ying [42] take inspiration from the peer-to-peer digital currency bitcoin [43] to propose a cloudlet-based multilateral resource exchange framework for mobile users. It is a system of small data centers that enables seamless transactions between users over the Internet bandwidth as a proof of concept. The cloudlet is a more scalable and low-cost cloud deployment paradigm aimed at using nearby compute and storage resources. It leverages the resources of idle personal servers of individual users, accessible via Wi-Fi. Access to the resources requires 802.1x authentication, from a server and a wireless router via Radius. 802.1X is an authentication protocol for securing a computer's access to a network that can be either wired (internal wired network or LAN) or wireless (internal wireless network or WLAN) [44].

Wu and Ying present an in-market resource exchange system for mobile users that acts as a virtual money system. It is suitable for other distributed systems. However, the storage service is inaccessible without bandwidth. According to the mapping by the Global Connectivity Index (GCI) [45], the gap in connectivity between countries is widening. This digital divide is also visible even within the so-called connected countries since access to the Internet is not the same depending on whether the user is in a rural or urban environment. A solution will have to be found, within the framework of the cloudlet, for the exchange of resources between users without Wi-Fi [42]. We summarize the needs for improvements in security mechanisms relative to the preceding mobile payments in Table 4.2.

4.4 THE PRIVACY PROBLEM

The privacy problem becomes increasingly important when considering the large number of participants in a mobile payment system as outlined in the scenario of local commerce (Figure 4.1). Data is often accessible to hardware providers (S1 for stakeholder 1), mobile network operators (MNO as S2), payment service providers, banks, card networks, and merchant personnel. Some systems do not have a reliable mechanism for protecting this data in transit or in storage; others provide partial protection. In this section, we will briefly describe the different classes of solutions.

4.4.1 Privacy with the Considered Technologies

In the article, "USSD-Architecture analysis, security threats, issues and enhancements," a threat of exposure of the user's confidential data is expressed [46]: they are displayed as such during the transaction. Moreover, for some mobile payment systems based on SMS and USSD, the data is encrypted with algorithms that have already undergone the process of reverse engineering [47] with the risk of

TABLE 4.2
Improvements Needed for Security in Mobile Payments

Types of Mobile Payment Systems	Safety Measures to Be Taken
• Payment system using NFC with SE integrated in the mobile.	• Avoid the limitations of the preregistered card.
• Payment system using NFC with HCE architecture.	• Strengthen the symmetric key to encrypt the exchanges and the IP and MAC addresses of the user's mobile. • Protect the mobile application against attacks from malicious programs that could access preloaded confidential data. • Secure the communication between the HCE server and the user's cell phone to avoid a relay attack.
• Systems using cryptographic primitives.	• Adopt cryptographic operations to ensure good performance and scalability.
• System based on distributed platforms, especially with cloudlet.	• Secure the links, as the risk has become greater in a distributed environment, as some nodes may be breached. • Overcome the lack of connectivity in some regions.

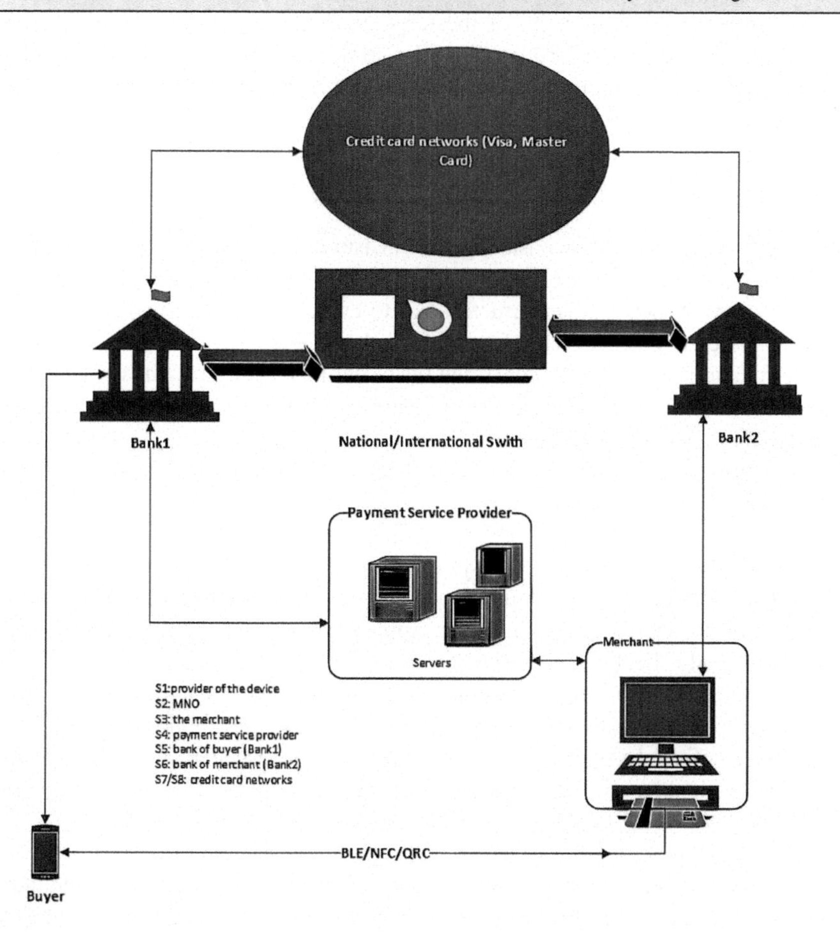

FIGURE 4.1 Payment process in a mobile payment system for local commerce scenario.

cloning the SIM that hosts the encryption key. Finally, it should be mentioned that end-to-end encryption is not guaranteed and that the mobile operator will have access to the data before reaching the service provider when the two entities are different.

For BLE and SQRC, cryptography protects the exchanges [24]. This prevents a large number of intermediaries from having access to sensitive data, but in this case, merchants still access payment card numbers. Furthermore, with BLE and QRC, sensitive data (especially private keys) can be stolen by malware on the customer's mobile. For BLE specifically, this data can be read by an attacker nearby while using the system. Some NFC architectures are more efficient using tokenization. Thus, the merchant is entitled to a token. However, in all cases, the service provider or bank has access to a set of information that is part of the customer's digital identity.

Another entity not considered in the diagram in Figure 4.1 can be referred to as the Third-Party Payment Provider (TPP) which is between the merchant and the user's bank [48]. PayPal is one of the most popular TPPs. The user's payments pass through these entities. Thus, in their activities, they must include mechanisms for the protection and security of the large amounts of data to which they have access. These are obligations made by existing laws. However, it is understood that increasing the number of intermediaries in this way also increases the risk of leakage of sensitive data.

4.4.2 Privacy with Biometric Authentication

Biometric authentication is basically the measurement and comparison of intrinsic biometric characteristics, specific to a person (fingerprint, face, iris, etc.) with a stored copy of these same characteristics to establish the match [49]. Often, biometric characteristics, extracted from the user's raw biological data, are stored with other personal information to constitute his or her identifier. Thus, he or she can use his or her biometric characteristics to authenticate himself or herself when accessing a payment service, for example.

The conventional operation of a biometric authentication system in mobile payment shows two main steps:

- First, the user enrolls in the system by providing the requested biometric characteristics. These are usually stored on the payment processing server. However, some architectures (like Apple Pay) store these characteristics on the mobile client's equipment.
- Upon authentication, the customer captures the biometric data with their mobile device and sends it to the server processing the mobile payment. The latter will apply algorithms to verify the match between the newly acquired biometric data and the stored one. This task is performed on the customer's mobile device if the biometric features are on the customer's side.

This authentication method is increasingly adopted in the mobile payment industry because it inspires more trust in the user and because it is very convenient [50].

4.4.2.1 Protection of Stored Biometric Data

While in the case of Apple Pay [51], biometric data is kept in an enclave, a Trusted Execution Environment (TEE) of the user's mobile [52], this is not the case for all systems where this data may be less secure. If kept, on the user's mobile, malware can steal this data; it can be read if the user is close to an attacker and his Bluetooth is activated [28]. The risk of leakage of this data is also real when stored on the server; the service provider must then be reliable and have robust mechanisms for protecting this data. Thus, one of the major issues with biometric authentication systems is the protection of stored biometric data to prevent their fraudulent use.

4.4.2.2 Protection during Exchanges with the Remote Server

Another issue to consider for biometric authentication is the protection of data sent to the server. This issue is present as soon as there is an exchange of identification or payment data. Cryptography (and in particular encryption) is a well-known solution for this problem [53]. However, this requires the use of adequate protocols, since we saw in section 2.1.1 that one of the most used protocols (SSL/TLS) contained flaws that have been revealed.

4.4.2.3 Biometric Data Protection Solutions

Protecting biometric data is of the utmost importance. Not only can this data be used by the attacker to obtain services by impersonating the authentic client (impersonation), but also a stolen biometric characteristic cannot be modified in the same way as a password. Many solutions are known for the protection of stored biometric data. In the following, we give a brief review of two main categories of solutions as presented by Iynkaran Natgunanathan et al. in the article "Protection of Privacy in Biometric Data" [50].

The first category is one that could be called "Biometric Encryption" and consists in encrypting the biometric data using a cryptographic technique. There are two operating modes in this category: Key Binding and key generation techniques from biometric data. In the first mode, a random secret key is homogeneously mixed with the biometric features using cryptographic tricks. The second mode is rather a set of techniques to build a cryptographic key from the biometric data and a sketch usually saved on a server. The second category is the technique known as "Cancelable biometric" or CB which consists in intentionally adding a distortion to protect the user's privacy [54].

4.4.2.4 Balance between Accuracy and Privacy

Moreover, to have more robust and accurate biometric authentication systems, more biometric features need to be extracted from the raw biological data and stored [55]. Feature extraction is sometimes done using machine learning methods on a large dataset. This clearly increases the risk of exposing private data. It has been shown that in some cases, a user's face can be reconstructed with the stored features to establish the match. This is then a violation of the intended irreversibility feature in biometric authentication systems.

4.5 PROTECTING PRIVACY AND ENHANCING DIGITAL TRUST

Privacy-related computing systems must be trusted to avoid security breaches, one of the sources of distrust in technology. This section describes four highly secure mobile payment system prototypes that are capable of improving users' digital trust in mobile payment systems. The first prototype is associated with TrustZone.

4.5.1 Secure Computer System and TrustZone

Authors Zheng, Yang, Shi, and Meng [56] propose a platform framework using a secure computing system for payment privacy through TrustZone-enabled platforms. TrustZone technology ensures the privacy of sensitive data from malware. Authors Zheng, Yang, Shi, and Meng use a mobile payment platform that is a prototype system in a simulation environment. They rely on Advanced RISC Machines and originally Acorn RISC Machine (ARM) architectures and virtualization and develop a system with FastModels which are accurate and flexible models of ARM IP [57]. An implementation is presented on a real development plan using ARM CoreTile Express A9x4. The platform can ensure the security of payment transactions, realize benevolent payment regarding privacy, and provide reliable computing services. It can also prevent malicious Robot Operating System (ROS) attacks and can secure the display and input to prevent the reading of sensitive data of the display device and input devices by hostile agents.

In order to improve the practicality of mobile payment, a touch screen, secure fingerprint recognition will need to be created. In addition, it will be necessary to improve the security of the payment mechanisms in a real environment [56]. Any mobile payment system needs to secure the data generated by the transactions to be robust. Kang and Nyang [58] respond to this need for security.

4.5.2 Transit Payment System and Privacy Protection

Kang and Nyang propose a privacy-protecting payment-in-transit prototype. The protection is based on traceable signatures, identity, and anonymous signatures. In addition to privacy, the system facilitates proactive blocking of misbehaving passengers. It supports free transfer services with postpaid programs allowing mobile payments using smartphones.

In the proposed system, transit agencies cannot obtain the identity of passengers and payment seekers cannot obtain passenger itineraries. Passengers can complete their entry and exit procedures in approximately 0.3–0.4 seconds, including revocation verification, which takes approximately 0.1 seconds. Although the proposed system is designed for transit services such as metro systems, it can be applied to offline mobile payment systems [58]. The transit payment system will need to be tested in different real-world environments to ensure its effectiveness. The third prototype of a particularly well-secured mobile payment system is presented in the following. It uses asymmetric encryption.

4.5.3 Asymmetric Encryption and Encrypted QR Code

Purnomo, Gondokaryono, and Kim [59] propose a mutual authentication technique between the customer and the merchant using the encrypted "Quick Response" (QR) code. The encrypted QR code allows data to be shared quickly, securely [60]. The secure transmission of the data transaction can be done by using an encrypted QR code. The mobile payment application requires a secure system to gain the digital trust of the customer or client. Security in the payment system is fundamental as it involves personal bank account data. To protect both the buyer and the seller, a higher level of security using mutual authentication is required to ensure the security of the transaction. Mutual authentication is achieved through a "Public Key Infrastructure" (PKI) system. The use of this PKI will ensure the security of the key distribution. This encryption system uses the RSA algorithm (named after the initials of its three inventors Ron Rivest, Adi Shamir, and Leonard Adleman), which is considered to be the most powerful message protection system [61]. This security system is also enhanced with the addition of the encrypted QR code as a payment medium. The use of the encrypted QR code will ensure the integrity, readability, and confidentiality of the information. It is also an important basis for the security of the system. However, if an intruder manages to steal the secret or private key, he will be able to access the information [59].

Asymmetric encryption will be able to serve as a model for application in the blockchain environment which is a secure and decentralized technology for storing and sharing data. It is currently at the heart of Estonia's e-service economy. It helps to secure digital identity, which is based on mobile technology or the use of smartphones for authentication and electronic signature. Four principles [62] define this digital society and can be used by public or private organizations for a transparent data management:

- decentralization, instead of a centralized database, each member of a network can create its own database;
- interconnection, a harmonization concerning all the members is achieved thanks to a free and open-source data link layer that is X-Road (https://x-road.global/);
- the openness of the platform, the PKI system is used; and
- the openness of the processes, the project is continuously developed and improved.

These four principles can be seen as the foundation for an ethical administration of big data that is not sensitive. They refer to the collaboration of an inclusive community (of members, for example), the openness of the source code, and the transparency of the history (not the content) of exchanges. In this sense, it is worth highlighting, on a global scale, the projects of the Linux Foundation [63] and the World Wide Web Consortium (W3C) [64] that support open-source systems and the development of open standardization. At the national level, countries can also promote these ethical principles, as Estonia, in particular, and

Europe, in general, has done through its GDPR. Considering the case of bitcoin – which is a free and open peer-to-peer digital currency with no central authority [43], the blockchain seems to be the technology that implements these ethical specifications, even if the ecological problem arises depending on the energy sources that power the servers.

Although bitcoin offers a solution for transparency without violating the security of user-generated data [65], questions of privacy remain for sensitive data like financial data.

4.5.4 Blockchain and Privacy

The question of privacy has often been raised in the context of the deployment of public blockchains such as bitcoin. Indeed, bitcoin offers a very large accessibility to the history of peer-to-peer transactions. Access to history is also meant to be the proof of transparent governance that information systems so badly need. However, this transparency could also prove problematic for the preservation of the privacy of participants. Especially since privacy is a requirement of some national laws or professions. Section 103 of Quebec's Law 25 is a good example of this privacy requirement. Anonymous communication networks provide answers to this problem. This is the case of Tor which "provides anonymous connections that are strongly resistant to both eavesdropping and traffic analysis" [66]. Private blockchains also offer privacy protection. This is why the recent EOSIOS proposal seems to us to be of great relevance. In May 2021, with the aim of addressing a need for privacy in the blockchain ecosystem was created "EOSIO data privacy working group" [67]. Contributors include Blockstalk, Block.one, dfuse, EOS Costa Rica, Digital Scarcity, Europechain, EOS Amsterdam, Rewired.One, and Gimly. Through the different deployments, five main themes [67] are identified:

- Use of cryptography is prevalent among blockchain deployments
- On-chain contracts facilitate coordination
- There is a wide range of code-sharing practices for transactions
- Innovations around data privacy would help accelerate data migration
- Shared ledgers upstage multiple blockchain-based approaches.

It is important to specify that another interest of a private blockchain is the possibility to make access to data inaccessible, which, according to Thibaut Labbé, allows to obtain a "result similar to a deletion" [68]. Furthermore, it would be wise to consider the option of Self Sovereign Identity [69], which gives the citizen the right to share or not share his data. In other words, a data user expert or wise person can choose to limit access to his or her data or even to permanently terminate access to his or her data. The definitive end to access to one's data can be considered as a form of destruction of one's data.

If the use of blockchain requires, most of the time, an Internet connection, the "contactless payment" (NFC) can be made offline, while being secured. This is what the fourth prototype developed by A. M. San and C. Sathitwiriyawong [70].

4.5.5 Contactless Payment and Privacy

The paper "Privacy-Preserving Offline Mobile Payment Protocol based on NFC" [70] considers a new offline mobile payment protocol based on NFC technology. In the system, a group signature is used to ensure the noncapability of transactions. Since there is less computation at the payment phase, the efficiency of the group signature is improved. The system is appropriately designed to meet all security and privacy requirements, including anonymity, non-mobility, randomness, and prevention of repeated attacks. It focuses on protecting the privacy of customers for mobile payment. First, the customer requests an anonymous ID from his or her bank. Then, the customer uses this ID to register with the Trusted Service Manager (TSM). The TSM only knows that the customer is authorized by the bank, but does not know the customer's real identity. When the customer generates a group signature for the transaction message, only the TSM knows the customer's anonymous identity. Therefore, the system protects the customer's privacy. The offline mobile payment protocol based on NFC by San and Sathitwiriyawong [70] can be implemented where the digital divide problem still exists.

The different cases that have been presented above clearly demonstrate that security remains a major issue in mobile payment. When the vulnerability of an already implemented system is known, the reputation of the company using the technology suffers. Security is an essential element in building digital trust. It can be enhanced both by the use of asymmetric encryption, with an Internet connection, and by contactless payment, in the case of offline mobile payment. Table 4.3 shows various proposals for solving security problems.

TABLE 4.3
Security Mechanisms in Mobile Payment

System Types and References	Mechanisms and Models for Protecting Privacy and Enhancing Digital Trust
Secure mobile payment with third-party platforms [57].	A so-called digital trust environment (TrustZone) is dedicated to the system to be secured, separating it from the external environment where all the applications are present.
Transit payment system with privacy protection [59].	The identity of the client is kept by the system. However, a participant (financial institution or transport company) only has access to the information he or she has provided. To achieve this, traceable and anonymous signatures are used.
Payment system using QRC with exchange encryption [60].	The system proposed in the reference uses QRC for communication between the seller's and buyer's systems. It is based on public key cryptography with mutual authentication of the interlocutors.
Blockchain and privacy.	A review of the literature on EOS and non-EOS blockchains that focus on privacy solutions.
Contactless payment with privacy protection [70].	The system uses group signature: the authorization for a client comes from his or her institution and is transmitted to a digital trust service manager.

4.6 ETHICAL RECOMMENDATIONS FOR PROTECTING THE LIFE CYCLE OF PERSONAL PAYMENT DATA

We have looked at several technologies that condition payment transactions. With each use of these technologies by an individual, multiple categories of data are generated as shown in section 2. This personal data has a lifecycle that evolves in four stages: the moment of collection, the moment of storage, the moment of exploitation, and the moment of destruction or anonymization (see Figure 4.2).

4.6.1 Collection

Collection can be manual or automatic. Personal payment data collection requires consent. In Europe, the contract is one of the legal bases under the GDPR [71]. In other words, where the collection is objectively necessary for the performance of the contract, consent is not required. However, secondary use of the payment data requires an explicit collection of consent [72].

4.6.2 Storage

Storage is the process of recording and preserving data using a server or cloud services. Given the value of payment data, it becomes relevant to know where the payment data of a country's citizens is stored. Even though in the case of a payment card in Canada, where applicable, we know that the data related to purchase is communicated to the "financial institutions and payment processing companies" [73], it remains unclear

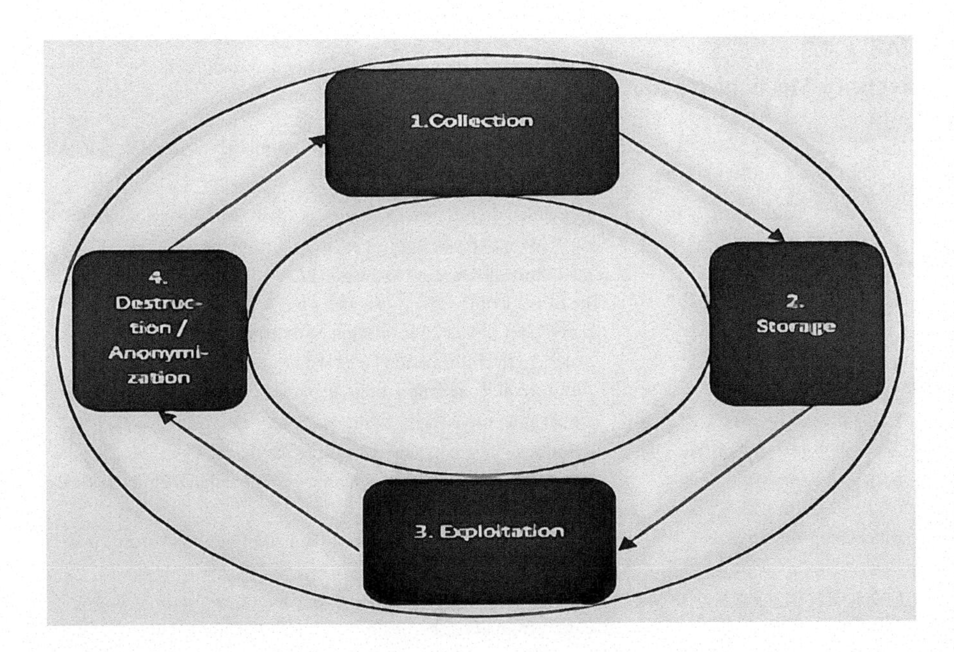

FIGURE 4.2 Data life cycle management.

where the collected information is stored. From a privacy perspective, it would be appropriate for countries to require storage in secure national servers and sovereign clouds. In the presence of biometric data, the CNIL suggests local storage in order to give full control to the individual [74]. It is tempting to use the services of American companies that have democratized cloud services. However, one should not forget that the Clarifying Lawful Overseas Use of Data (CLOUD) Act [75] could be an invasion of the privacy of data of citizens of any country. A U.S.-based global provider must facilitate access to the data it stores if the U.S. deems it necessary.

4.6.3 Exploitation

The use of personal data can take many forms: use for the customer who must pay, use for the merchant who sells a service or product, secondary use of the data, sharing of the data, transfer to another country, etc.

4.6.4 Destruction or Anonymization

When the purpose for which the data was collected or used is achieved, it is important that the data be destroyed. A retention period must be considered in order to ensure audits (taking into account the requirements of professional bodies) or the guarantee of a product or service. According to Law 25, personal data can be anonymized, if there is a serious and legitimate reason (119). The anonymization of payment data could benefit the open data sector. Municipalities, cities, and countries can put anonymized citizen data to work for citizens. It is not right that payment data benefit, in large part or only, financial institutions.

Mobile payments raise many other ethical issues that have not been considered in this chapter: the risks of bias associated with parameter choices for establishing credit ratings or profiles; the new privacy issues arising from open banking; authentication and accountability related to the environment of connected objects with or without human intervention; privacy in light of the legalization of bitcoin; the new risks associated with peer-to-peer payment applications; and the eco-responsibility dimension.

4.7 CONCLUSION

This literature review presents some characteristics of mobile payment technologies and systems. USSD, BLE, SMS, WAP, QRC, and NFC/RFID are the models described in the article. Most mobile payment systems implemented around the world use one or more of these technologies. Many of them raise ethical issues that concern respect for personal data. Ethics concerns all mobile payment stakeholders: users, researchers, entrepreneurs, the public sector, privacy associations, nonprofit organizations, and many others. So mobile payment is not just for entrepreneurs or the private sector. It is important that all stakeholders in the ecosystem participate in public consultations and the implementation of privacy policies for the benefit of mobile payment users. As mobile has established itself as an accessible tool, it is becoming easier to access private data and track the online activity of its users. This is why security is a major priority in the mobile payment field. In this sense, asymmetric encryption is an excellent vector. It meets the need for data security, especially in this period of strong growth in mobile payment usage. Given the magnitude of the problems that have been described in the previous sections, it is an ethical choice that can be combined with blockchain for the development of highly secure mobile payment systems.

NOTE

1. This chapter is an extension of a previous article published in French in 2021. Pierre, Schallum et Italis, Olson (2021). « Les systèmes de paiement mobile à l'ère de la COVID-19: sécurité, vie privée et confiance numérique », vol 6 L'innovation collaborative, revue *Technologie et innovation*, ISTE OpenScience. DOI : 10.21494/ISTE.OP.2021.0598

REFERENCES

[1] Samsukha, A., "Mobile Payment Statistics & Facts 2022 for Marketers," *Emizentech*, November 7. https://www.emizentech.com/blog/mobile-payment-statistics-facts.html, 2021.

[2] Market Research Report, "Digital Market Payment," *Fortune Business Insights*, January, 2020. https://www.fortunebusinessinsights.com/digital-payment-market-101972, 2020.

[3] Académie de la transformation numérique (ATN) de l'Université Laval, "Services bancaires en ligne," *NETendances*, édition 2021, volume 12 – numéro, p. 17. https://api.transformation-numerique.ulaval.ca/storage/57/netendances-2021-services-bancaires-en-ligne.pdf.

[4] Xiao Y., Fan Z., « 10 tendances technologiques à surveiller pendant la pandémie de COVID-19, » 2020. https://fr.weforum.org/agenda/2020/05/10-tendances-technologiques-a-surveiller-pendant-la-pandemie-de-covid-19/.

[5] Pymnts, "European Data Protection Supervisor Urges Increased Data Protection in Card Payments," January 4, 2022. https://www.pymnts.com/news/security-and-risk/2022/european-data-protection-supervisor-urges-increased-data-protection-in-card-payments/, 2022.

[6] Jones D., « COVID-19 Pandemic Caused Increase in Cyber Fraud and Changes in Banking, » *Atm Marketplace,* June 19, 2020. https://www.atmmarketplace.com/articles/cyber-fraud-surges-as-covid-19-changes-banking-e-commerce-2/.

[7] Canadian Centre for Cyber Security, *Staying Cyber-Healthy during COVID-19 Isolation*, 2020. https://www.cyber.gc.ca/en/news-events/staying-cyber-healthy-during-covid-19-isolation.

[8] Columbus, L., « 2020 Roundup of Cybersecurity Forecasts and Market Estimates » Forbes, Editors' Pick, 2020. https://www.forbes.com/sites/louiscolumbus/2020/04/05/2020-roundup-of-cybersecurity-forecasts-and-market-estimates/#39fc8ba6381d.

[9] PCI Security Standards Council, *Securing the Future of Payments: PCI SSC Publishes PCI Data Security Standard v4.0*, March 31, 2022. https://www.pcisecuritystandards.org/about_us/press_releases/securing-the-future-of-payments-pci-ssc-publishes-pci-data-security-standard-v4-0/.

[10] Justice Laws Website, Personal Information Protection and Electronic Documents Act (S.C. 2000, c. 5), September 28, 2022. https://laws-lois.justice.gc.ca/ENG/ACTS/P-8.6/index.html.

[11] Justice Laws Website, "Canadian Payments Act (R.S.C., 1985, c. C-21)," 2022. https://laws-lois.justice.gc.ca/eng/acts/c-21/.

[12] Assemblée nationale du Québec, "Bill 64, An Act to modernize legislative provisions as regards the protection of personal information," http://assnat.qc.ca/en/travaux-parlementaires/projets-loi/projet-loi-64-42-1.html, 2021.
[13] [Gouvernement du Québec, *Ministère de la Cybersécurité et du Numérique*, 2022. https://www.quebec.ca/gouvernement/ministere/cybersecurite-numerique/mission-mandats.
[14] Eur-Lex, «General Data Protection Regulation», *Official Journal of the European Union*, 2019. https://eur-lex.europa.eu/legal-content/EN/TXT/?uri=CELEX:32016R0679.
[15] Agence nationale de la sécurité des systèmes d'information (ANSSI), 2020. https://www.ssi.gouv.fr/.
[16] Commission Nationale de l'Informatique et des Libertés (CNIL), 2020. https://www.cnil.fr/.
[17] Sorensen, E., "Different Types of Mobile Payments Explained," *Mobile transaction*, May 31, 2018. https://www.mobiletransaction.org/different-types-of-mobile-payments/.
[18] Lerner, T., « Mobile Technology and Security », dans T. Lerner (dir.), *Mobile payment*, Springer Vieweg, Mainz, 2013.
[19] Pathirana, P. A., Azan S. M. F., « Factors Influencing the Use of Mobile Payments — A Conceptual Model », dans *2017 National Information Technology Conference (NITC)*, Colombo, Sri Lanka, 2017.
[20] Italis, O., Étude comparative des plateformes de paiement mobile, Mémoire de maîtrise, Institut des sciences, des technologies et des études avancées d'Haïti, 2018.
[21] Gupta, N. K., *Inside Bluetooth Low Energy, second Edition*, Artech House Publishers, Boston, 2016.
[22] Padgette, J., Bahr, J., Batra, M., Holtmann, M., Smithbey, R., Chen, L., Scarfone, K., « Guide to Bluetooth Security, » 2017. https://nvlpubs.nist.gov/nistpubs/SpecialPublications/NIST.SP.800-121r2.pdf.
[23] Pierre, S., « Introduction à la mobilité et aux systèmes cellulaires » dans S. Pierre (dir.), *Réseaux et Systèmes Informatiques Mobiles: Fondements, Architectures et Applications*, Presses internationales Polytechnique, Montréal, 2011.
[24] Sherif, M. T., « Mobile Payment » dans M. T. Sherif (dir.), *Protocols for Secure Electronic Commerce*, CRC Press, Boca Raton, 2016.
[25] Castéran, S., « Le paiement sans contact n'est pas sans risque, » 2015. https://lejournal.cnrs.fr/articles/le-paiement-sans-contact-nest-pas-sans-risque.
[26] CNIL, When Trust Pays Off: Today's and Tomorrow's Means of Payment Methods Facing the Challenge of Data Protection," 2022, p. 10. https://www.cnil.fr/sites/default/files/atoms/files/cnil-white-paper_when-trust-pays-off.pdf.
[27] Pierre, S., « La vie privée à l'heure du Big Data et la « minimisation des données », » *Alternatives économiques*, 2023. https://blogs.alternatives-economiques.fr/reseauinnovation/2023/03/26/la-vie-privee-a-l-heure-du-big-data-et-la-minimisation-des-donnees.
[28] Wang, Y., Hahn, C., Sutrave, K., « Mobile Payment Security, Threats, and Challenges », dans *2016 Second International Conference on Mobile and Secure Services (MobiSecServ)*, Gainesville, FL, 2016.
[29] Szurdi, J., Chen, Z., Starov, O., McCabe, A., and Duan, R., « Studying How Cybercriminals Prey on the COVID-19 Pandemic», *Unit 42*, April 22,, 2020. 2020.https://unit42.paloaltonetworks.com/how-cybercriminals-prey-on-the-covid-19-pandemic/.
[30] Europol, « Rapport: Pandemic Profiteering: How Criminals Exploit the COVID-19 crisis, », 2020. https://www.europol.europa.eu/publications-documents/pandemic-profiteering-how-criminals-exploit-covid-19-crisis.
[31] Digicert, « Qu'est-ce qu'un certificat SSL? » 2020. https://www.websecurity.digicert.com/fr/ca/security-topics/what-is-ssl-tls-https.
[32] Torres, G., « Attaques de l'homme du milieu: qu'est-ce que c'est et comment les éviter, » 2018. https://www.avg.com/fr/signal/man-in-the-middle-attack/.

[33] QYR-13533967, « Global Near Field Communication Market Size, Status and Forecast 2019–2025 » 360 Market Updates, 2019 https://www.360marketupdates.com/global-near-field-communication-market-13533967.

[34] Black, T., « Votre chèque est à la poste? Vivement que cette phrase disparaisse! Paiements Canada, » 2020. https://www.paiements.ca/votre-cheque-est-la-poste-vivement-que-cette-phrase-disparaisse.

[35] Pasquet, M., Gerbaix, S., « Fraud on Host Card Emulation Architecture: Is It Possible to Fraud a Payment Transaction Realized by a Mobile Phone Using an "Host Card Emulation" System of Security? », dans 2016 Second International Conference on Mobile and Secure Services (MobiSecServ), Gainesville, FL, 2016.

[36] Thales, « Sécurité des paiements sans contact: la technologie HCE trace la voie », 2015. https://www.thalesgroup.com/fr/systemes-dinformation-critiques-et-cybersecurite/event/securite-des-paiements-sans-contact-la.

[37] Paessler, « Qu'est-ce qu'une adresse IP? » 2020. https://www.fr.paessler.com/it-explained/ip-address.

[38] Till Halbach, T., « A prototype-based case study of secure mobile payments », dans *eChallenges e-2015 Conference: IEEE*, Vilnius, Lithuania, 2015.

[39] Antoni, J.-P., Vuidel, G., « MobiSim: un modèle multi-agents et multi-scalaire pour simuler les mobilités urbaines », dans J.-P. ANTONI, *Modéliser la ville. Forme urbaine et politiques de transport*, Economica, coll. Méthodes et approches, 2010.

[40] Urien, P., « Innovative Mobile Payments in the Cloud for Connected Citizen: The Mobisim Project », dans *18th Mediterranean Electrotechnical Conference (Melecon): IEEE*, Lemesos, Cyprus, 2016.

[41] Yeh, K. H., « A Secure Transaction Scheme with Certificateless Cryptographic Primitives for IoT-Based Mobile Payments », *IEEE Systems Journal* 12, no. 2, pp. 2027–2038, 2018.

[42] Wu, Y., Ying, L., « A Cloudlet-Based Multi-lateral Resource Exchange Framework for Mobile Users », dans *2015 IEEE Conference on Computer Communications (INFOCOM)*, Kowloon, 2015.

[43] bitcoin.org. https://bitcoin.org. 2020.

[44] Gonzales, L., « Qu'est-ce que le 802.1X? » 2013. https://blog.devensys.com/introduction-authentification-reseau-802-1x/.

[45] Global Connectivity Index (GCI), « Harnessing the Power of Connectivity: Mapping Your Transformation into a Digital Economy with GCI 2017, » 2017. https://www.huawei.com/minisite/gci/assets/files/gci_2017_whitepaper_en.pdf?v=20191217v2.

[46] Lakshmi, K. K., Gupta, H., Ranjan, J., « USSD—Architecture Analysis, Security Threats, Issues and Enhancements », dans *2017 International Conference on Infocom Technologies and Unmanned Systems (Trends and Future Directions) (ICTUS)*, Dubai, 2017.

[47] Nyamtiga, B. W., Anael, S., Loserian, S. L. « Security Perspectives for USSD Versus SMS in Conducting Mobile Transactions: A Case Study of Tanzania, » *International Journal of Technology Enhancements and Emerging Engineering Research* 1, no. 3, pp. 38–43, 2013.

[48] Valcke, P., Vandezande, N., Van de Velde, N, "The Evolution of Third Party Payment Providers and Cryptocurrencies under the EU's Upcoming PSD2 and AMLD4" (September 23, 2015). SWIFT Institute Working Paper No. 2015-001, Available at SSRN: https://ssrn.com/abstract=2665973

[49] Ogbanufe, O., Kim, D.-J., Comparing Fingerprint-Based Biometrics Authentication Versus Traditional Authentication Methods for Epayment, *Decision Support Systems*, vol. 106, pp. 1–14, 2018

[50] Natgunanathan, I., Mehmood, A., Xiang, Y., Beliakov, G., Yearwood, J., "Protection of Privacy in Biometric Data," in *IEEE Access*, vol. 4, pp. 880–892, 2016, doi: 10.1109/ACCESS.2016.2535120.
[51] Apple Inc. 2020. Face ID security. https://support.apple.com/en-us/102381.
[52] Margraf, M., Lange, S., Otterbein, F., « Security evaluation of apple pay at point-of-sale terminals », dans *2016 10th International Conference on Next Generation Mobile Applications, Security and Technologies (NGMAST)*, Cardiff, 2016.
[53] Wang, Feng, Shan, Ge Bao, Chen, Yong, et al., "Identity authentication security management in mobile payment systems," *Journal of Global Information Management (JGIM)*, 2020, vol. 28, no 1, pp. 189–203.
[54] Ratha, N. K., Connell, J. H., Bolle, R. M., "Enhancing security and privacy in Biometrics-Based Authentication Systems," *IBM Systems Journal*, vol. 40, no. 3, pp. 614–634, April 2001.
[55] Osadchy, M., Dunkelman, O., « It Is All in the System's Parameters: Privacy and Security Issues in Transforming Biometric Raw Data into Binary Strings, » *IEEE Transactions on Dependable and Secure Computing* 16, no 5, p. 796–804, 2019.
[56] Zheng, X., Yang, L., Shi, G., Meng, D., « Secure Mobile Payment Employing Trusted Computing on TrustZone Enabled Platforms, » dans *2016 IEEE Trustcom/BigDataSE/ISPA*, Tianjin, 2016.
[57] ARM Developer, « Fast Models. ». 2020. https://developer.arm.com/tools-and-software/simulation-models/fast-models/.
[58] Kang, J., Nyang, D., A Privacy-Preserving Mobile Payment System for Mass Transit », *IEEE Transactions on Intelligent Transportation Systems* 18, no. 8, p. 2192–2205, 2017.
[59] Purmono, A.T., Gondokaryono, Y.S., Kim, C.S., « Mutual Authentication in Securing Mobile Payment System Using Encrypted QR Code Based on Public Key Infrastructure », dans *2016 6th International Conference on System Engineering and Technology (ICSET)*, Bandung, 2016.
[60] Ahamed, M. S., Mustafa, H. A., « A Secure QR Code System for Sharing Personal Confidential Information », dans *2019 International Conference on Computer, Communication, Chemical, Materials and Electronic Engineering (IC4ME2)*, Rajshahi, 2019.
[61] Asaduzzaman, A., Gummadi, D., Waichal, P., « A Promising Parallel Algorithm to Manage the RSA Decryption Complexity », dans SoutheastCon 2015, Fort Lauderdale, 2015.
[62] Murphy, A., « Estonia's Mobile-ID: Driving Today's e-Services Economy », *https://www.gsma.com/identity/wp-content/uploads/2013/07/GSMA-Mobile-Identity_Estonia_Case_Study_June-2013.pdf.* 2013.
[63] The Linux Foundation, 2020. https://www.linuxfoundation.org/.
[64] World Wide Web Consortium. 2020. https://www.w3.org/.
[65] Khan, B., Syed, T., « Recent Progress in Blockchain in Public Finance and Taxation », dans *2019 8th International Conference on Information and Communication Technologies (ICICT)*, Karachi, 2019.
[66] Wikipedia, Tor, 2019. https://en.bitcoin.it/wiki/Tor.
[67] EOSIO, "The State of Blockchain Privacy," May 13, 2021. https://eos.io/news/blockchain-privacy/?utm_source=twitter&utm_medium=social&utm_campaign=corporate_news&utm_content=EOSIO_data_privacy.
[68] Labbé, T., Le droit face aux technologies disruptives: le cas de la blockchain, p. 192. Droit. Université de Strasbourg; CEIPI. Français, 2021. https://tel.archives-ouvertes.fr/tel-03518287/document.
[69] Strüker, J., Urbach, N., Guggenberger, T., Lautenschlager, J., Ruhland, N., Schlatt, V., Sedlmeir, J., Stoetzer, J.-C., Völter, F., "Self-Sovereign Identity Foundations, Applications, and Potentials of Portable Digital Identities," *Fraunhofer Institute for*

Applied Information Technology FIT Project Group Business & Information Systems Engineering, 2021. https://www.fit.fraunhofer.de/content/dam/fit/de/documents/Fraunhofer%20FIT_SSI_Whitepaper_EN.pdf.

[70] San, A. M., Sathitwiriyawong, C., « Privacy-Preserving Offline Mobile Payment Protocol Based on NFC », dans 2016 International Computer Science and Engineering Conference (ICSEC), Chiang Mai, 2016.

[71] CNIL, Le contrat: dans quels cas fonder un traitement sur cette base légale?, 2020. https://www.cnil.fr/fr/les-bases-legales/contrat.

[72] CNIL, When Trust Pays Off: Today's and Tomorrow's Means of Payment Methods Facing the Challenge of Data Protection," 2022, p. 59. Op. Cit.

[73] Office of the Privacy Commissioner of Canada, "Electronic and Digital Payments and Privacy," 2016. https://www.priv.gc.ca/en/privacy-topics/technology/mobile-and-digital-devices/02_05_d_68_dp/.

[74] CNIL, When Trust Pays Off: Today's and Tomorrow's Means of Payment Methods Facing the Challenge of Data Protection," 2022, p. 58. Op. Cit.

[75] The United States Department of Justice, "CLOUD Act Resources," 2022. https://www.justice.gov/criminal-oia/cloud-act-resources.

Section III

Sensitive Data Challenges

5 Where Does the Novel Legal Framework for AI in Canada Stand against the Emerging Trend of Online Test Proctoring?

Céline Castets-Renard
Civil Law Faculty, University of Ottawa, Ottawa, Canada

Simon Robichaud-Durand
Master of Law and Technology, Law Faculty,
University of Ottawa, Ottawa, Canada

CONTENTS

5.1 Introduction 98
5.1.1 AI Discrimination 98
5.1.2 Socio-Economic Discrimination 98
5.1.3 Privacy 100
5.1.4 Variety of Technologies in Education 100
5.1.5 Exam Surveillance Tools and Pandemic 100
5.1.6 Economic Value of the Education Technology Market 101
5.1.7 AI and Privacy 101
5.2 Overview of Monitoring Software and the Role of AI 101
5.2.1 Overview of the Main Tools 101
5.2.2 Role of AI 102
5.2.3 Functions of AI Systems 102
5.2.4 Risks of Errors and Discrimination 103
5.2.5 Hybrid Systems and Human Control 103
5.3 Risks Generated by AI Tools for Exam Monitoring 103
5.3.1 Main Legal Issues 103
5.3.2 Provincial Laws Regarding Personal Information in the Public Sector 104
5.3.3 Provincial Laws on Personal Information in the Private Sector 104
5.3.4 Liability in the Public–Private Partnerships 105
5.3.5 Nature of Personal Data Collected 106

DOI: 10.1201/9781003227656-9

5.3.6 Applicable Laws and Location of Personal Data Storage 106
5.3.7 Obligations in Some Provinces to Store Personal Data.................... 107
5.3.8 Opacity of AI Tools 108
5.4 AI Framework in Canada: Federal Bill C-27 (Part III) and Risk Minimization 108
5.4.1 Privacy and AI Reform.. 108
5.4.2 Limited Data Protection Reform 109
5.4.3 Legal Framework of AI 109
5.4.4 Definition of AI System.. 109
5.4.5 Territorial Scope 110
5.4.6 Material Scope.. 110
5.4.7 High-Impact Systems... 110
5.4.8 Person Responsible 111
5.4.9 Data Anonymization.. 111
5.4.10 High-Impact Mitigation 112
5.4.11 Transparency Measures 112
5.4.12 Responsibility 112
5.5 Recommendations to the Federal Legislature for the Adoption of Future Regulations 114
5.5.1 Definition 114
5.5.2 Reinforce Transparency... 115
5.5.3 Error and Bias Risk Attenuation Must Be Better Specified 116
5.5.4 Responsibility Chain.. 116

5.1 INTRODUCTION

5.1.1 AI Discrimination

Generally, the conversations surrounding risks or harms deriving from the use of AI have often been dominated by the overreach of power such as public surveillance or police surveillance using AI facial recognition software, with a potential for discrimination, such as race, gender, and age biases. Several examples of Uighur minorities surveillance in China by Huawei[1] or in the West by police forces using Clearview AI tools[2] have been widely publicized. They show us their risk for societal harm and the need for ethical principles.

5.1.2 Socio-Economic Discrimination

Though AI proctoring software is not immune to the main types of discrimination mentioned above, there are additional layers one must consider when it comes to the potential for discrimination in this type of software. By design, proctoring software flags potential instances of cheating and, in attempting to do so, it also can pick up other regular occurrences and flag them as suspicious activity that it deems as an attempt to cheat. Such activities could include the audible voice of a person or a domestic pet, the presence of a second person on a camera, even if they are just passing by, the lighting quality, and the internet connection quality.

This type of risk is related to the fact that students take exams off-campus, most often at home where they are potentially not alone or quiet. Students may then be placed in an inequitable position for socio-economic reasons. The use of proctoring software by universities adds new layers to potential social discrimination, that are often secondary thoughts in the discussions of discriminations by AI; granted recently talks surrounding COVID-19 contact tracing apps have begun to tackle this type of discrimination. Therefore, the use of AI-based proctoring software is a particularly important and distinctive scenario, that regulations must consider. As stated earlier, some examples of non-conventional factors leading to possible discrimination can pertain to some basic and fundamental life choices, such as having a domestic pet or not, having children or living with children, residing in a multigenerational household, your physical location (urban or rural), and your accessibility to required equipment such as a computer, a webcam, lighting, or the possibility of occupying a private and quiet space.

One's decision of owning or caring for a domestic pet is regarded as a personal choice that doesn't often lead to a disadvantageous position, albeit landlord discrimination towards pet owners has long existed. In the case of online proctoring software, having a domestic pet or being in the presence of one during an exam could trigger a software alert. More precisely, a domestic pet who makes noise, such as a bark or a chirp, during an online proctoring exam has been identified as a cause for flagging a potential cheating incident. So much so that ProctorU mentioned this in their 2021 communication[3] regarding their move towards using less automated systems. Granted this potential risk is not limited to animals, as the same risk and occurrence has happened with children. The presence of children speaking in the background has set proctoring systems to flag this as a potential cheating behaviour, as confirmed in ProctorU 2021 communication. Considering that many students own domestic pets or are parents, the potential for harm is not insignificant. Elsewhere, due to the fact that online proctoring software requires test takers to be in an isolated room that is quiet and well lit, the potential for harm to students living in large or multigenerational homes is also prevalent. Additionally, due to material requirements to use such online proctoring software, such as a computer, a webcam, adequate lighting, and an internet connection, this creates another possible discrimination factor. Furthermore, as a whole, Canada boasts a high internet access rate; 94% of Canadians have access to the internet. However, high-speed internet access (a connection of at least 50mMbps download and 10Mbps upload) in rural Canada is much more limited; the CRTC estimated that only 53.4% of rural Canadian communities have access to high speeds.[4] Therefore, students living in rural areas are also faced with another burden that could flag an online proctoring software if the quality of their internet connection was too low or if their connection dropped.

Lastly, because individuals could realistically belong to a number of these living scenarios, the potential for discrimination is also compounded. Therefore, hypothetically, a student located at intersectionality[5] who belongs to a more known discrimination group, such as race, gender, or age, who has or lives with a family pet, or a child, who resides in a multigenerational family home, and who is in a rural town, would be subjected to an extreme risk of discrimination by online proctoring software.

5.1.3 Privacy

On the other hand, while the online surveillance takes place in a private place where the test is taken, there is no doubt that this process invades privacy, as an Ohio judge has explicitly recognized.[6] Federal Judge J. Philip Calabrese determined that the university's room scan did constitute an "unreasonable search." "Mr. Ogletree's subjective expectation of privacy at issue is one that society views as reasonable and that lies at the core of the Fourth Amendment's protections against governmental intrusion."

5.1.4 Variety of Technologies in Education

The objectives motivating the use of technology in the educational field vary greatly.[7] In some instances, technology serves to assist the admission selection process for universities or other reputable schools;[8] in other instances, it shares resources on online platforms such as a virtual campus.[9] Additionally, technology has also been used to evaluate the academic performance of students[10] or attribute a final grade to students.[11] Therefore, academic surveillance can be considered as an emerging field of "capitalism surveillance" pertaining to the dominance of a few companies in the surveillance field.[12] Elsewhere, the use of video conference tools such as Zoom has also been prevailing during strict containment periods of the COVID-19 pandemic,[13] which, in turn, came at the expense of student's privacy and their cybersecurity.[14]

5.1.5 Exam Surveillance Tools and Pandemic

Among the many technological usages in the educational field, our focus will solely be pertaining to online proctoring software used by Canadian universities. The COVID-19 pandemic forced many to adapt, such as working remotely, and the academic world was not spared. Thus, Canadian universities had to pivot quickly and opted to offer remote learning, which took the shape of online classes and exams, especially during the strict containment periods.[15] According to certain studies, students are more tempted to cheat during an online exam compared to an in-person exam.[16] Granted, occurrences of cheating aren't limited to the online environment; however, virtual exams made it easier to cheat and more tempting than before, all the while students are dealing with a heighten level of stress.[17] Therefore, it's said cheating occurrences have become more widespread during online exams when comparing them to their in-person counterparts.[18] Furthermore, this tendency of online cheating has been observed worldwide.[19] Thus, it's not surprising that universities began to adopt online proctoring software, following claims that the use of this software could attenuate cheating instances during online exams.[20] However, student bodies from all over the world expressed their worries regarding this type of surveillance, for instance in Australia,[21] in the United States,[22] and in Canada, as seen at the University of Ottawa[23] and the University of Manitoba.[24] Further proof of this is in the United States a website has been created to track which American universities are using online proctoring software.[25]

5.1.6 Economic Value of the Education Technology Market

Historically, online proctoring software isn't a novelty. In Canada, this type of software has been used in the educational field for over 20 years.[26] For example, Queens University has been using this type of software for many years.[27] Before the COVID-19 pandemic, Ed-Tech was a blooming sector and, by the end of 2019, that sector was valued at an estimated 19 billion USD.[28] By 2020, the larger market size of the online learning sector was valued at an estimated 100 billion USD.[29] Furthermore, it is projected that from 2020 to 2027, the market size pertaining solely to online proctoring software will grow 16.4% annually[30] and in 2027 it should represent an estimated value of 10 billon USD.[31] Certain authors highlighted a 720% growth in the usage of online proctoring software since the beginning of the 2020 pandemic.[32] It is clear that online proctoring solutions are developing into a profitable market and some authors warn that the uptake in the implementation of these solutions could potentially be used as a proof of concept to justify the deployment of online proctoring software.[33]

5.1.7 AI and Privacy

Exam proctoring software mobilizes a significant amount of student personal information that is used to enable AI systems to perform academic integrity checks, but also to train the AI systems. This is done while the consent is general and not specific to these different uses.

This paper will explore some of the socio-economic related risks, including risks for privacy, surrounding the use of AI-assisted proctoring software by Canadian universities. To do so, Section 2 will focus on the various software that has been deployed in Canadian universities and the role that AI plays to identify potential cheating incidents. In Section 3, we will highlight the risks generated by using proctoring software utilizing AI, followed by Section 4, where we will examine Bill C-27, the emerging legal framework in Canada. Lastly, in Section 5, we will present eight recommendations to the federal legislator regarding future regulations in the Canadian IA landscape that would better protect Canadians

5.2 OVERVIEW OF MONITORING SOFTWARE AND THE ROLE OF AI

5.2.1 Overview of the Main Tools

Online proctoring software represent a variety of tools often based on artificial intelligence, such as Respondus Monitor, Proctorio, ProctorU, ProctorExam, Examity, and ProctorTrack. The six online proctoring software can vary in their level of intrusiveness on the student's privacy. Very often, the online proctoring software is integrated in a university's Online Learning Platform or DSL and maintained by a third-party company for a university, as is the case with the University of Ottawa's use of Respondus Monitor[34] via Brightspace's Virtual Campus,[35] with the University of Manitoba,[36] and with Ryerson University.[37] To start, Respondus Monitor is an online proctoring software who states that it offers a one-stop shop for online proctoring tailored to higher education.[38] Proctorio, another online proctoring software, has been used by Laval University[39] and Concordia University.[40] Third is ProctorU,[41] another online proctoring software that has seen an uptake by the University of Toronto,[42] University of Waterloo,[43] Athabasca University,[44] Thompson Rivers University,[45] and Memorial University.[46] ProctorExam[47] is another online proctoring software that is used by Polytechnique Montréal to accommodate students requests to write an exam virtually.[48] Examity is another online proctoring software that offers a live solution, which is said to be done by highly qualified individuals using a separate software to supervise the exam in order to maintain the highest degree of academic integrity.[49] In Canada, Examity is used by the University of Toronto.[50] Finally, ProctorTrack is another online proctoring software company who offers various solutions. In Ontario, ProctorTrack has been selected by the province's e-campus; programme and licences to ProctorTrack software are funded by the province and made available to academic institutions.[51] Universities such as Queen's University[52] and Western University[53] have used ProctorTrack for their online proctoring needs. Elsewhere, in Saskatchewan, the University of Regina used ProctorTrack.[54] However, due to a breach of security, the company has stopped offering their services.[55] These few examples help paint the portrait of the use of online proctoring software by universities in Canada, motivated by the need to adapt to the new standard of online learning with technology-driven solutions.

5.2.2 Role of AI

The presence of AI in proctoring software is somewhat of a novelty. During the 2010s, proctoring software was highly dependent on human intervention, notably human proctors who would supervise students in real time. However, in hopes to improve precision and efficiency of online proctoring, today's software implements artificial intelligence software.[56] Thus, today's proctoring software now require less human proctoring in order to identify exam takers and detect cheating instances due to their algorithmic dependence.[57]

5.2.3 Functions of AI Systems

The features of online proctoring software vary; some of them only lock the exam taker's screen, while others record students via video and analyse the recordings using AI to detect instances of cheating. For instance, Respondus Monitor is an online proctoring solution that is entirely automated which requires students to be equipped with a webcam to record the length of their online exam. Following the exam, the potential cheating instances are then flagged and are provided to the instructor for verification.[58] Sometimes, the video recording functions are associated to AI systems such as facial recognition or automated surveillance to track behaviours that the system deems to be abnormal. These behaviours can vary, but in the case of software using facial recognition technology, the disappearance of a subject's face or the appearance of a second subject can trigger the report of a potential cheating instance. Furthermore, the software solutions using facial recognition technology utilize biometric markers to identify students when they're logging into their exam and flag potential cheating incidents during the exam.[59] Additionally, biometric keystroke analysis which serves to track keystroke data, eye tracking which monitors and analyses eye movements, audio monitoring which records and monitors students sonically, and facial detection are all methods that are used by some proctoring software.[60] Automated proctoring software will then notify the exam supervisor of potential cheating incidents or behaviour it deems to be abnormal for a subsequent human verification and confirmation of the reported incident.[61]

Generally, proctoring software is tasked with identifying and authenticating the student, limiting the functionality of the student's computer, analysing student behaviour, and generating a report.[62] In the case of Proctorio, their software is equipped with features such as automated authentication, automated surveillance, video and audio recording, room recording, and the analysis of the students' behaviours.[63]

5.2.4 Risks of Errors and Discrimination

Proctorio states that their algorithmic techniques are superior and more effective than human proctoring and could fill in the gaps for human error, thus resulting in better results than human proctoring when detecting for cheating incidents.[64] However, many times, Proctorio has been accused of utilizing algorithms that are discriminatory towards visible minorities.[65] Based on some findings, it's been reported that 57% of authentication attempts by student with darker skin tones were unsuccessful.[66] Furthermore, it's also been said that the best authentication rate for all ethnicity groups is inferior to 75%, which highlights that the error rate is quite significant.[67]

5.2.5 Hybrid Systems and Human Control

It seems like universities are growing suspicious regarding fully automated systems and therefore are prioritizing hybrid systems. Hybrid systems are ones that combine human proctoring and automated proctoring which integrates artificial intelligence.[68] The Okanagan University of British Columbia (UBC) has since communicated that it will prohibit the use of automated online proctoring software such as Proctorio, in all of their programmes with the exception of accredited programmes who require such use.[69] In regard to Proctorio, the software has been subject to a Privacy Impact Assessment by UBC and the unsatisfactory results alone could justify its abandonment.[70] In the case of ProctorU, it offers many hybrid solutions such as Review+, which utilizes human proctors for the authentication and the post-verification of online exams, but the surveillance of the exam is automated.[71] ProctorU's Record+ option utilizes an automated system for the authentication of students and the surveillance of the exam, and results are then certified by a human proctor prior to emitting an incident report.[72]

5.3 RISKS GENERATED BY AI TOOLS FOR EXAM MONITORING

5.3.1 Main Legal Issues

Online proctoring solutions have raised ethical and legal issues. Some of these issues are regarding privacy violations, while others are pertaining to the protection of personal information,[73] and there are also concerns in regard to fundamental rights (equality and non-discrimination) in the implementation of AI solutions.

5.3.2 Provincial Laws Regarding Personal Information in the Public Sector

The use of proctoring software implies that some personal information is being collected in an academic setting. Those data are collected and used by academic institutions and the collection is justified by their provincial educational mission, and they are located in different provinces; therefore, provincial privacy laws are applicable. Academic institutions are public entities and the provincial public legal framework pertaining to personal information is applicable. The control factor of the *Privacy Act*[74] clarifies this division of jurisdiction and consequently the scope of the provincial framework. It applies to personal information and documents collected by public institutions such as universities. Therefore, the provincial public sector framework applies to student and university relationships.

However, when speaking of the control factor, control does not need to be absolute, direct, or permanent. In this regard, in *Canada Post Corp. v. Canada (Minister of Public Works)*,[75] the federal court of appeal deemed that the term "under the control" must be given a broad and liberal interpretation. For instance, if personal information is held by a representative or a service provider of a public institution, that information then pertains to the institution and, therefore, the legal framework regarding the public sector would apply. Given that universities often rely on third-party service providers, this precision is paramount when considering the use of online proctoring software. However, the above-mentioned case law is not applicable per se. Indeed, universities are not only deferring the collection and management of personal information to third parties, but they also put students and third-party companies in close cooperation. Furthermore, those third-party companies are also engaged in data collection practices pertaining to personal data that the universities do not have access to and cannot control. Granted, in this particular situation, on one hand the third-party companies are acting for public institutions, but, on the other hand, they are seeking to collect students' personal information for means that are entirely commercial, often using technological tools and, thus, the universities cannot prevent them from doing so. If universities wish to impose terms pertaining to the protection of personal information inspired by the provincial private sector framework and their privacy policies, they are then confronted with the reality of technological tools that they do not fully understand nor master. Not only is this observation pertinent to the collection of data, but it is also pertinent when considering the use and the storing of students' personal information.

5.3.3 Provincial Laws on Personal Information in the Private Sector

Should we consider that the provincial legal framework regarding the protection of personal information in the private sector applies? Many factors point towards the implementation of those laws in student and tech companies' relationships. First, the terms and conditions or the privacy policies state that personal information can be used for machine learning activities to allow companies to improve the performance of their systems. Such a finality surpasses what is foreseen by universities when considering online exam proctoring activities. Elsewhere, the collect of personal information is subject to students consent which is obtained prior to the use of the software. While possibilities for refusal are most often provided, students are in fact put in the difficulty of refusing because it could deprive them of having the same exam as other students (breach of equality) and requires them to openly declare their refusal (breach of anonymity). The originating mission justifying the collection of personal information is linked to the universities' mission of academic integrity and thus authorized by the provincial laws of the public sector without requiring consent as stated in the legal framework of the *Privacy Act*. However, the need for consent is then justified by the fact that personal information will be used for purely commercial means. This change in finality for which the personal information was collected surpasses its legal authorization, and therefore it justifies the need for consent. This change in finality is all the easier given that the personal information can be used to train various types of algorithms for a variety of usages. In the case of facial recognition technology, the same algorithms used for identifying students could be used for policing means, or the surveillance of targeted populations. The finality here is public in regard to the missions on public institutions; however, at the same time they are also commercial in regard to tech companies.

5.3.4 Liability in the Public–Private Partnerships

To better understand the different uses of personal information and data controllers, we must specify which role the universities and the tech companies play. For instance, ProctorU's privacy policy states that "[i]n most processing contexts, Meazure acts as a processor or service provider for a controller educational institution or certifying entity. In those situations, we process personal data only on documented legally-compliant instructions from the controller entity."[76] However, that is not the case when you consider the tech companies using personal information to improve their systems or better train their facial recognition algorithm. In such an instance, the universities lose the control of how students' personal information is used and, furthermore, they cannot be held liable. In theory, tech companies should be liable under PIPEDA, the federal legal framework pertaining to the private sector, except for Québec, Alberta, and British Columbia who have their own legal framework applicable to the private sector. However, certain companies claim to respect the applicable legislation. For instance, ProctorExam is a Dutch company subjected to the GDPR's regulation. Polytechnique Montréal and ProctorExam reached an agreement pertaining to the respect of the GDPR's regulation and Quebec's *Act respecting access to documents held by public bodies and the protection of personal information*.[77] In the case of the University of Toronto, an agreement was made with Examity who claims to protect the privacy of recordings and other personal information.

One thing for sure, most software is offered by private actors, who are outside of the public education mission of universities. Consequently, the distinction between private sector and public sector activities is lost and represents a profound limit considering that the current Canadian legal framework, both provincial and federal, relies on this distinction. Due to the numerous private-public partnerships, who are all too often misunderstood by the legislators, the protection of personal information weakens.

5.3.5 Nature of Personal Data Collected

Students are often asked to give up information when engaging with education software. In addition to identification information such as surname, name, and student number, directly given by students themselves, several other information is taken or can be deducted directly from the tech companies. For instance, Respondus Monitor[78] notes that they treat data taken from the exam session such as the date, time in which a student started and finished their exam; the exact time when a student responded to a question; the time spent on each question; if and when an answer has been modified; the quality of the student's internet connection during the exam, which includes if the students connection dropped; the activity of the student's mouse, their keyboard, and their screen; the quality of the video recording which consists of the lighting, the contrast, and the movements; and the audio recording's quality. This behaviour data constitutes personal information. Furthermore, there are times when online proctoring software utilizing AI collects biometric data, which is also considered as sensitive data, to identify a student's identity. This type of use would indicate the presence of a database of students' pictures in order to match with a live person recorded by a webcam. Lastly, the list of personal information that is subject to collection is clearly enumerated, which is usually contrary to one's right to information. Additionally, when reading the contractual terms, the minimization principle, the necessity principle, and the proportionality principle do not seem to be respected. Therefore, there is a real risk of hyper-surveillance for students.

5.3.6 Applicable Laws and Location of Personal Data Storage

Elsewhere, given that most service providers are located in the United States and that surveillance data is usually transferred to them, the application of the federal data protection legal framework is questionable and, if applicable, is the protection effective and sufficient? Most American companies contractually impose the transfer of data to the United States. Most of the time, cloud services such as Amazons Web Service (AWS) are used for storing data. Respondus' privacy policy indicates that if students are located elsewhere then where the servers are located, students' data is subject to be transferred internationally. This also applies when support is given from another country. The Ottawa University's website state that personal information can be stored outside of Canada and would therefore be subject to the laws where stored, which is in accordance to Ontario's *Freedom of Information and Protection of Privacy Act*. In Respondus' privacy policy, it is stated that the company uses AWS servers to store its data, which are located in the United States.[79] As per the contractual conditions of Respondus,[80] any legal questions arising pertaining to Respondus Monitor or its use are subject to the legislative interpretation of state of Washington, omitting to consider the conflict of laws.

ProctorExam, used by the Polytechnique Montréal, is a Dutch company subject to the General Data Protection Regulation (GDPR) and its regime. On its website, the company specifies who could potentially have access to their data.[81] It specifies that personnel authorized by the universities along with some of the ProctorExam's staff mandated by universities could have access to some data. It's worth noticing that ProctorExam also calls upon other companies to offer their services. However, ProctorExam states that it remains responsible for their third-party data processing. In the case of ProctorExam, their third-party partners are Amazon Web Services who stores data physically in the European Union (EU) and Google Cloud who also stores data in centres located in the EU. The EU storing obligation deriving from EU law which requires that that data is stored on the European territory is met here.

However, ProctorExam also states that "[i]f your institution has consented, we may also use third-party solutions to process selected data."[82] Therefore, Polytechnique Montréal's website state the company which ProctorExam uses for proctoring and review purposes is based in the United States. It's clear that data moves from Europe to Canada and then from Canada to the United States. The international data that flows from Europe is subjected to rules pertaining to the adequacy of data protection of Section 41 of the GDPR, but nothing is stated in regard to the respect of those legal obligations. It is not clear what is the legal basis for the transfer. This could suppose the respect of contractual obligations, such as model contractual clauses adopted by the European Commission to guarantee a minimum level of protection.

Examity,[83] used by the University of Toronto,[84] states that "Examity's Site is hosted and operated in the United States ('US'). Unless you have been informed otherwise, by using the Site, you are consenting to the transfer of your personal information to the US. If you are accessing the Site from outside the US, please be advised that US law may not offer the same privacy protections as the law of your jurisdiction."[85] Furthermore, the Examity's website redirects readers to a document titled "European Union or the European Economic Area, please see our European Union and European Economic Area Resident Privacy Notice"[86] which mentions the bilateral agreement between the European Union and the United States, also known as the Privacy Shield, invalidated by the European Court of Justice in 2020![87]

5.3.7 Obligations in Some Provinces to Store Personal Data

It should be highlighted that additionally to the fact that the Ontario's act regarding the public sector does not have obligations to store personal information on Canadian territory, other provinces have legislated such obligations such as British Columbia[88] and Nova Scotia.[89] Alberta[90] imposes an obligation of informing when the data is stored outside of Canada. In Québec, the *An Act to modernize legislative provisions as regards the protection of personal information*, also known as act 25, adopted in September of 2021[91] states that:

> "[b]efore releasing personal information outside Québec, a public body must conduct a privacy impact assessment. The body must, in particular, take into account
>
> (1) the sensitivity of the information; (2) the purposes for which it is to be used; (3) the protection measures, including those that are contractual, that would apply to it; and (4) the legal framework applicable in the State in which the information would be released, including the personal information protection principles applicable in that State.
>
> The information may be released if the assessment establishes that it would receive adequate protection, in particular in light of generally recognized principles regarding the

> protection of personal information. The release of the information must be the subject of a written agreement that takes into account, in particular, the results of the assessment and, if applicable, the terms agreed on to mitigate the risks identified in the assessment.
>
> The same applies where the public body entrusts a person or body outside Québec with the task of collecting, using, releasing or keeping such information on its behalf."[92]

Laval University, who utilizes Proctorio, communicated to students that their data is stored in Canada.[93]

Elsewhere, even if PIPEDA, the federal law applicable to the private sector, does not prohibit data from being stored outside of Canada, the Privacy Commissioner of Canada has repeatedly imposed an obligation of informing Canadians as to the location of where their data is being stored.[94]

5.3.8 Opacity of AI Tools

Lastly, another risk is surrounding the use of AI systems. Tech companies that are offering their services as solutions to online proctoring utilizing AI systems can vary from data mining techniques to facial recognition systems. However, the use of these tools is often not clearly stated. For instance, Proctorio states that their algorithmic techniques are superior and more effective than human proctoring and are capable of filling in the gaps for human shortfalls and error to detect cheating instances. They also state that their software is capable of improving its accuracy over time, which indicates the presence of machine learning without clearly specifying it. Tech companies involved in developing or offering online proctoring software should be more transparent regarding the products they offer and the methods they use. Furthermore, the term online proctoring software encompasses a variety of different tools and techniques which all vary regarding their societal impact.

Online proctoring software developers lack transparency, which results in a limited knowledge of the software's functionality and, thus, a lack of control on the software by the universities.[95] Therefore, how online proctoring software operates remains unknown, which should be deeply concerning for universities who must maintain a certain control on tests and exams, along with the methods used to ensure academic integrity. The lack of transparency is particularly concerning the nature and the quantity of personal information collected by online proctoring software, as is with the implementation and characteristics of IA systems.

5.4 AI FRAMEWORK IN CANADA: FEDERAL BILL C-27 (PART III) AND RISK MINIMIZATION

5.4.1 Privacy and AI Reform

The following section will highlight the novel Bill C-27 and part III, which we deem to be pertinent in the legal analysis of online proctoring software. In doing so, we've opted to focus on the legal framework of artificial intelligence (AI), the definition of AI system, the territorial scope, the material scope, high-impact systems, the notion of responsible person, data anonymization, high-impact mitigation, transparency measures, and responsibility all pertaining to the *Artificial Intelligence and Data Act* of Bill C-27 (part III).

5.4.2 Limited Data Protection Reform

On June 16, 2022, the House of Commons of Canada unveiled Bill C-27, which is comprised of three sections, titled *an Act to enact the Consumer Privacy Protection Act, the Personal Information and Data Protection Tribunal Act and the Artificial Intelligence and Data Act and to make consequential and related amendments to other Acts.*[96] The first law serves as a reform to the *Personal Information Protection and Electronic Documents Act* (PIPEDA). The previous weaknesses raised, such as the lack of clarity on the liability of private-public partners, the quantity and the nature of data collected, the location of out of country data storing, as well as the applicable laws, are still recurring challenges when considering the use of technological tools. However, these issues are not considered in Bill C-27, and therefore will remain when considering the issues of data protection of students.

5.4.3 Legal Framework of AI

The third law is the new *Artificial Intelligence and Data Act* (AIDA)[97] and is the most applicable regarding online proctoring software. Granted, the use of AI in society has only grown over the years and todays AI systems are deployed in many different sectors of society, thus forcing governments to start legislating the use of AI.

5.4.4 Definition of AI System

An AI system means a: "technological system that, autonomously or partly autonomously, processes data related to human activities through the use of a genetic algorithm, a neural network, machine learning or another technique in order to generate content or make decisions, recommendations or predictions."[98]

This definition is not without recalling the proposed AI act by the European Commission of April 21, 2021 regarding the use of AI systems.[99] Section 3(1) of the AI act states that an: "'artificial intelligence system' (AI system) means software that is developed with one or more of the techniques and approaches listed in Annex I and can, for a given set of human-defined objectives, generate outputs such as content, predictions, recommendations, or decisions influencing the environments they interact with."[100] We can clearly notice that the primary objectives assigned to those systems are identical, as they pertain to generating content, making recommendations or predictions, and making decisions.

As for determining the categories of techniques or the methods used, AIDA can seem limited by nature given that it refers to genetic algorithms, generative adversarial networks, and machine learning. However, it does state "or another technique," meaning that the techniques aimed by AIDA are not exhaustive and allow for a larger interpretation of evolving technologies. The European Commission's proposal for regulation lists the technologies that are subject to the regulation in Appendix 1, using a broad approach since it encompasses machine learning,[101] approaches based on logic and knowledge,[102] and approaches bases on statistics.[103]

5.4.5 Territorial Scope

This Act does not apply with respect to a government institution as defined in Section 3 of the Privacy Act. Section 4 states that "the purposes of this Act are (a) to regulate international and interprovincial trade and commerce in artificial intelligence systems by establishing common requirements, applicable across Canada, for the design, development and use of those systems; and (b) to prohibit certain conduct in relation to artificial intelligence systems that may result in serious harm to individuals or harm to their interests."[104]

As noted by Professor Teresa Scassa, "regulating the digital economy has posed some constitutional (division of powers) challenges for the federal government, and these challenges are evident in the AIDA, particularly with respect to the scope of application of the law. [...] By focusing on international and interprovincial trade and commerce, the government asserts its general trade and commerce jurisdiction, without treading on the toes of the provinces, who remain responsible for intra-provincial activities."[105]

Online proctoring software have been designed and developed for private companies, most often American ones, and on the contrary have vocation to be governed by this future law. We must then be satisfied that the territorial scope is broad and covers international trade. Canada should then be able to impose its regulation on companies located abroad who engage in selling their systems to Canadian business' or to government entities.

5.4.6 Material Scope

Section 5 of AIDA specifies the substantive scope when defining a "regulated activity" as: "any of the following activities carried out in the course of international or interprovincial trade and commerce: (a) processing or making available for use any data relating to human activities for the purpose of designing, developing or using an artificial intelligence system; (b) designing, developing or making available for use an artificial intelligence system or managing its operations."[106]

American companies allowing their AI systems to be available, such as an online proctoring software, clearly fit the territorial and substantive scope of the law.

5.4.7 High-Impact Systems

It is moreover necessary to consider the "high impact" of the AI system on the population to which it will be applied. According to article 5(2): "a high-impact system means an artificial intelligence system that meets the criteria for a high-impact system that are established in regulations."[107]

To mitigate the risks generated by high-impact systems, several requirements for the responsible party are enacted throughout Sections 6 to 12. Section 7 first requires an assessment of whether an AI system is one of high impact: "a person who is responsible for an artificial intelligence system must, in accordance with the regulations, assess whether it is a high-impact system."[108]

Such regulation has yet to be adopted. Therefore, it is still too early to confirm if online proctoring software will be considered as high-risk systems subject to additional requirements. It is important to note here that the law plays little role in favour of regulation. The legislator delegates to the government the power to set the norms on a technical subject that is very important in all aspects of the lives of Canadians, which is highly questionable. As Professor Teresa Scassa says: "this crucial term in the Bill will mean what cabinet decides it will mean at some future date. It is difficult to fully assess the significance or impact of this statute without any sense of how this term will be defined."[109]

Faced with these uncertainties and given the previous issues pertaining to their error rates and discrimination, it would be preferable that systems with the highest degree of automation are classified as high-impact systems. Either way, we propose this hypothesis as the point of departure from our analysis, to consider the requirements that should be implemented to the responsible authorities of AI systems.

5.4.8 Person Responsible

In addition, Section 5(2) of AIDA defines a responsible person as "a person is responsible for an artificial intelligence system, including a high-impact system, if, in the course of international or interprovincial trade and commerce, they design, develop or make available for use the artificial intelligence system or manage its operation."[110]

We must recognize the broad approach in the qualification of the accountable authority that can very well be the developers or the distributers. The multiple accountable parties create a chain of accountability, but it is however desirable to determine the roles of each to specify the rules regarding the division of liability based on each party's defined role. To guarantee the implementation of the liability principals, further regulation should bring clarification to this issue.

5.4.9 Data Anonymization

Section 6 of AIDA sets out special rules relating to the protection of personal information by requiring compliance with data anonymization requirements. It states that: "a person who carries out any regulated activity and who processes or makes available for use anonymized data in the course of that activity must, in accordance with the regulations, establish measures with respect to (a) the manner in which data is anonymized; and (b) the use or management of anonymized data."[111]

When applied to online proctoring software, the requirements of data anonymization of personal information would be quite relevant. We must reiterate that online proctoring software collects an important amount of personal information such as biometric data which is highly sensitive data, as is with behavioural data of students in a highly stressful environment, which could lead to a student's intellectual capacities being inferred and the profiling of individuals. Such aim would clearly surpass the mission of maintaining academic integrity. Yet, when considering that this data can also be used to improve the performance of, and capacity of, AI systems, the goal to improve efficiency cannot lead to a violation of individual protection. Therefore, data anonymization is not only an important means to consolidate performance and protection, but it must also be encouraged.

5.4.10 High-Impact Mitigation

Section 8 of AIDA states that: "a person who is responsible for a high-impact system must, in accordance with the regulations, establish measures to identify, assess and mitigate the risks of harm or biased output that could result from the use of the system."[112] Moreover, "a person who is responsible for a high-impact system must, in accordance with the regulations, establish measures to monitor compliance with the mitigation measures they are required to establish under section 8 and the effectiveness of those mitigation measures."[113]

It is then important that the regulation clearly states the means pertaining to minimizing risks.

5.4.11 Transparency Measures

On another note, Articles 10 to 12 of AIDA set out transparency measures. Section 10(1) states measure for keeping general records: "a person who carries out any regulated activity must, in accordance with the regulations, keep records describing in general terms, as the case may be, (a) the measures they establish under sections 6, 8 and 9; and (b) the reasons supporting their assessment under section 7."[114] Moreover, "the person must, in accordance with the regulations, keep any other records in respect of the requirements under sections 6 to 9 that apply to them."[115]

5.4.12 Responsibility

Section 11 of AIDA places the responsibility for publishing the description of AI systems on the person responsible for the system (1) and on the person who manages the operation of the system (2). In the case of exam proctoring software, this implies that both the companies providing the systems and the universities using them would be required to publish information describing them. More precisely:

> *(1) A person who makes available for use a high-impact system must, in the time and manner that may be prescribed by regulation, publish on a publicly available website a plain-language description of the system that includes an explanation of (a) how the system is intended to be used; (b) the types of content that it is intended to generate and the decisions, recommendations or predictions that it is intended to make; (c) the mitigation measures established under section 8 in respect of it; and (d) any other information that may be prescribed by regulation.*
>
> *(2) A person who manages the operation of a high impact system must, in the time and manner that may be prescribed by regulation, publish on a publicly available website a plain-language description of the system that includes an explanation of (a) how the system is used; (b) the types of content that it generates and the decisions,*

> *recommendations or predictions that it makes; (c) the mitigation measures established under section 8 in respect of it; and (d) any other information that may be prescribed by regulation.*[116]

In the two instances, the information provided is of general nature and does not explain and assists with understanding how decisions are taken regarding individuals. Therefore, on the macro-scale, transparency should then be required. Furthermore, it is also asked for high-impact attenuation measures to be taken and that the public should be informed of said measures, as it is not guaranteed that the measures taken would be sufficient to prevent or alleviate the error and discrimination risks. Regarding profiling of individuals and the consequences that could arise, nothing is mentioned. The generality and the vague nature of the requirements fall short in bringing the necessary guaranties to combat the greater societal risks. Once again, it will be up to the government to specify all these aspects and parliamentarians do not have the opportunity to determine the rules of liability and the means of mitigating harms.

Finally, in the event of a result that could cause harm, Section 12 of AIDA foresees a duty of notifying the Minister in charge. In this regard, Section 12 states that:

> *[a] person who is responsible for a high-impact system must, in accordance with the regulations and as soon as feasible, notify the Minister if the use of the system results or is likely to result in material harm.*[117]

The Minister in charge of the implementation of the regulation also has the power to make arrangements to ask the responsible authority of an AI system to provide information regarding their system and its implementation (Sections 13 to 17). AIDA sets out governance rules that centralize power in the hands of the Minister, reinforcing the impression that the government wants to retain control over standard setting and implementation. As Mardi Witzel notes: "A single ministry, ISED, is proposed as the de facto regulator for AI in terms of law *and* policy making *and* administration *and* enforcement."[118] Moreover, the AIDA state that "the Minister may designate a senior official of the department over which the Minister presides to be called the Artificial Intelligence and Data Commissioner, whose role is to assist the Minister in the administration and enforcement of this Part." We agree with Mardi Witzel to say that "there is no independence from ISED or separation of roles."[119] Moreover, ISED has a crucial role as Canada's AI industry rapidly expands while there is a growing need for transparency in decision-making, accountability in decisions and access. This dual role of consumer protection and industry development may pose a tricky balance for ISED[120] and one can be surprised by the choice of a single ministry to represent divergent interests.

Lastly, many observations can be made pertaining to the implementation of AIDA. First, as we currently wait for the additional regulation, it is impossible to know specifically what they will be. Many provisions are left blank and have yet to be completed. For instance, we have no details about the main provisions of AIDA regarding: "biased output" from an AI system; "high-impact system"; assessment of AI system impact; and definition of material harm. Thus, it is impossible to know if they will be sufficiently pertinent and if they will be up to the task of addressing the ethical and legal risk of AI. We encourage the legislator to clarify these concepts during the parliamentary debate.

Consequently, it's not certain the risk surrounding online proctoring software – which is what concerns us here – will be considered as high-impact systems and thus in the scope of this novel law. Our postulate was one reflecting that this would be the outcome but, to this day, there is no certainty of this. Therefore, this must be a focal point to any upcoming regulation on the matter. Additionally, the duty of transparency foreseen in AIDA cannot, by itself, tackle the risk of error nor can it tackle bias which are generally found in the deployment of AI systems. They are insufficient in nature and, therefore, further regulation is needed and is much anticipated, as is with enacting of minimization measures.

5.5 RECOMMENDATIONS TO THE FEDERAL LEGISLATURE FOR THE ADOPTION OF FUTURE REGULATIONS

To contribute to the accuracy of the upcoming provisions of the regulation to come, several recommendations must be made to the legislator.

5.5.1 Definition

"High-impact" system must be defined. AIDA has an objective of regulating AI systems which introduce a "high impact," and this classification must satisfy the criteria established by the regulation. Two remarks must then be made.

First, the Canadian legislator seems to want to follow an inclusive approach based on individual risk factors or, more generally, societal risk factors. Conversely, the European Commission made the choice of listing certain use cases in Appendices II[121] and III[122] of the proposal for AI regulation. Listing the risks represents a normative precision advantage but a disadvantage regarding its limitation effect on the apprehension of risk, given that certain risks won't be specified. Evolutionary measures are made possible allowing the European Commission to review the list but that will of course cause a delayed process. Canada has chosen to go forward with a method that allows avoiding the possible risk of its legislation becoming obsolesce, but in doing so, at the risk of being imprecise. Therefore, we recommend that the Canadian legislator be more precise in their selection of criteria to facilitate their application rather than their interpretation. To do so, it could be useful to follow Directive on Automated Decision-Making developed by the Treasury Board of Canada in 2019.[123] Although it is non-binding and only reserved for decisions made by the federal government, this directive also considers measures to minimize the risks of automated decisions based on the incidence level. It also established criteria to determine those levels, considering the right of persons and communities; the health and well-being of persons and communities; the financial interest of individuals; and an ecosystem's sustainability. Subject to a few adaptations pertaining to the administrative sector and the private sector, those criteria could be useful to elaborate regulation.

Second, beyond proctoring software, numerous AI systems can have a "high impact." We should then encourage the Canadian legislator to adopt a broad approach to include the maximum number of situations and consider the risks of opacity, of error, of bias, and of discrimination. Elsewhere, we also suggest that "impact" be defined as "risk," including potential harm, not only actual harm.[124]

Recommendation 1 There is a need to define precisely the criteria that qualify "high-incidence" AI systems.

Recommendation 2 There is a need for a broad interpretation of the notion of "high impact," to consider the plurality of societal risks linked to security, health, and fundamental rights (opacity, error, and bias), including actual and potential harm.

5.5.2 Reinforce Transparency

AIDA lays down rules of transparency regarding the use of AI systems. As shown with online proctoring software, such a requirement is quite necessary, and must allow for the quantity of personal information collected and the nature of which automated processes are applied. That information is fundamental to understand the eventual risks of automation regarding errors or bias. This information also allows for the full exercise of subsequent rights. For instance, the right to give a personal explanation of a decision made, and the right to challenge, cannot be made possible without such information regarding the AI system and its function. Likewise, a human-controlled environment implies that a person has knowledge of the AI system to understand it's automation and be capable of evaluating it.

To guarantee such transparency, Section 10 of AIDA requires book keeping of general terms, the measures taken to abide to legal requirements, as well as the motives of evaluation of a "high-impact" system, the measures taken to attenuate the risks and to ensure its respect. It's also expected that a person who carries out regular activity pertaining to a high-impact system must publish its description on a publicly accessible website, using general terms, regarding its intended use, the content it will generate, the predictions, recommendations, or decisions it should be taking, along with the risk attenuation measures. Future regulation should also impose an obligation to provide further information. Granted, describing the intended use and the desired results is necessary but not sufficient to achieve a level of transparency which would guarantee an explanation, as well as a right to challenge and a right to recourse to individuals.

In contrast, the European Commission's proposal for regulation on AI foresees measures that are more specific that should be considered when drafting the upcoming regulations for the AIDA. Section 13 of the Proposal for regulation on AI pertains to transparency and the obligation to provide information to users in regard to "the characteristics, capabilities and limitations of performance of the high-risk AI system."[125] More specifically, the level of accuracy must be specified along with the level of robustness and cybersecurity "against which the high-risk AI system has been tested and validated."[126] Furthermore, Section 13 b (iv) also requires for the "performance as regards the persons or groups of persons on which the system is intended to be used."[127] Elsewhere, Section 14 of the Proposal for regulation on AI requires human oversight regarding the design and development of a high-risk AI system to allow effective control during the time the system is deployed. Such human control aims to avoid or minimize risks associated to health, security, or fundamental human rights. Finally, Section 15 of the Proposal for regulation on AI foresees provisions regarding the accuracy, robustness, and cybersecurity of AI systems who are supposed to "perform consistently in those respects throughout their lifecycle."[128] High-risk AI systems must also be resilient when it comes to errors, faults, or inconsistencies. In this regard, some of those measures are comparable in the Treasury Board of Canada's Directive on Automated Decision-Making.

Recommendation 3: It should be reminded that transparency measures must be specified, and they must require further and more precise information regarding high-incident AI systems pertaining to their features, capacities, and performance limitations, in considering vulnerable communities.

Recommendation 4: A reliable and effective human oversight should be imposed to prevent, or greatly reduce, the risks associated with health, security, and fundamental human rights.

Recommendation 5: Measures regarding resilience and evolution control systems must be imposed during the entire life cycle of an AI system.

5.5.3 Error and Bias Risk Attenuation Must Be Better Specified

AIDA considers measures of risk minimization pertaining to high-impact AI systems. The liable authority of a high-impact system must establish measures aimed at defining, evaluating, and attenuating risks and harms or biases that could result from the use of AI systems. Regulations must then specify the terms of its obligations.

Section 10 of the EU's Proposal for regulation on AI could also serve as inspiration to define the provisions regarding data and its governance. Training data sets, along with validation data sets and testing data sets, are subject to practices, for instance design choices, data collection, or processing operations related to data preparation such as annotation, labelling cleaning, enrichment, and aggregation. Data sets must be subjected to pre-assessments of data in regard to availability, quantity, and necessary adaptation of data sets, along with an assessment in order to reveal potential data biases, shortcomings, or deficiencies and the way they could be resolved. Training data sets, validation data sets, and testing data sets must be pertinent, representative, complete, and without errors. Section 15(5) also requires that high-impact AI systems who continue their learning after their market entry, or after they're put into service, for biases occurring from the utilization of errored data as entry data for future operation (feedback loops) can be lessened by appropriate attenuation measures.

It is probable that the Canadian legislator is not as interested in detailing measures as the European counterpart is. However, the control of data sets is still very much useful and needed. This kind of provisions can also be enacted in code of conducts or professional rules.

Recommendation 6: There is a need to consider the conditions of data set creation.

5.5.4 Responsibility Chain

Lastly, AIDA places the liability on certain actors, specifically the designers, the developers, and broadcasters, such as the vendors, distributors, or importers. We must highlight that there seems to be a desire to broadly regroup liable parties.

However, rules pertaining to the degree of liability of every party should be laid out. The rules deriving from Sections 16 to 29 of the EU's regulation proposal on AI could also serve as inspiration. Those rules set out in the proposal take aim at AI system providers, importers, distributors, and users which could all be liable when disregarding a rule. These different parties have different obligations.

Recommendation 7: Rules regarding shared liability should be specified for parties targeted in AIDA.

Finally, the Canadian legislator has yet to prohibit the use of certain AI systems; by contrast, the European Commission prohibits the use of four specific AI systems clearly defined at Section 5(1) of the Regulation proposal on AI.[129] It is therefore crucial to engage in a collective discussion to determine if certain use of AI systems should be prohibited due to their potential harms to individuals or society, such as lethal weapons or the general use of facial recognition in public spaces for surveillance purposes.

Recommendation 8: A collective reflection should be initiated on whether to prohibit certain uses of AI and the means to determine how to identify such prohibited uses.

In closing, it is important to keep up with the advancements of AIDA and the complementary rules to come. When considering online proctoring software, it is clear that these systems have been deployed in a context of weak legal requirements and their use highlights specific risks. This being simply one of many instances that should be considering when the Canadian legal framework intends to identify risk-based uses.

In conclusion, it is still too early to know what Bill C-27 and its part III (AIDA) will become, as parliamentary discussions are ongoing. However, the example of online test proctoring illustrates the dire need to strengthen the law in order to provide greater protection for those whom AI systems already apply and those who will be subjected to AI's widespread uptake.

NOTES

1. In Xinjiang, China, surveillance technology is used to help the state control its citizens: <https://www.cbc.ca/passionateeye/features/in-xinjiang-china-surveillance-technology-is-used-to-help-the-state-control>.
2. The American company Clearview AI offered its facial recognition services to Canadian police forces, including the RCMP, using a database of more than three billion images taken from the Internet without the users' consent. The joint investigation by Canadian privacy authorities concluded in June 2021 that there was a violation of provincial and federal privacy laws, but this serious breach did not result in any sanctions: Joint investigation of Clearview AI, Inc. by the Office of the Privacy Commissioner of Canada, the Commission d'accès à l'information du Québec, the Information and Privacy Commissioner for British Columbia, and the Information Privacy Commissioner of Alberta: <https://www.priv.gc.ca/en/opc-actions-and-decisions/investigations/investigations-into-businesses/2021/pipeda-2021-001/>.
3. Meazure Learning, "ProctorU to discontinue exam integrity services that rely exclusively on AI," May 24, 2021, online: <https://www.meazurelearning.com/resources/proctoru-to-discontinue-exam-integrity-services-that-rely-exclusively-on-ai>.
4. Canadian Radio-television and Telecommunications Commission (CRTC), "Internet," August 4, 2022, online: <https://crtc.gc.ca/eng/internet/internet.htm>.

5. Kimberle Crenshaw, “Demarginalizing the Intersection of Race and Sex: A Black Feminist Critique of Antidiscrimination Doctrine, Feminist Theory and Antiracist Politics,” *University of Chicago Legal Forum,* (1989) 1:1989.
6. *US District Court Northern District of Ohio, Eastern Division.* Aaron M. Ogletree v. Cleveland University, Case No. 1:21-cv-00500, August 22, 2022, online: <https://www.theverge.com/2022/8/23/23318067/cleveland-state-university-online-proctoring-decision-room-scan>.
7. Hannah Natanson, “Live vs. Tape-Delayed: How Two Approaches to Online Learning Change Life for Teachers and Students,” *Washington Post,* April 28, 2020, online: <https://www.washingtonpost.com/local/education/live-vs-tape-delayed-how-two-approaches-to-online-learning-change-life-for-teachers-and-students/2020/04/25/250fb7d0-7bfe-11ea-9bee-c5bf9d2e3288_story.html>.
8. For instance, in France, the “ParcourSup” platform to access higher education institutions, online: <https://www.parcoursup.fr/>.
9. Such as BrightSpace, which collects personal information and information about users' activity.
10. Predictive AI systems have been deployed to predict academic performance on cancelled assessments: see Simon Coghlan, Tim Miller et Jeannie Paterson, “Good Proctor or ‘Big Brother’? AI Ethics and Online Exam Supervision Technologies,” (2020), p. 3, online: <http://arxiv.org/abs/2011.07647v1>.
11. See, in the United Kingdom, the scandal generated by the use during the pandemic of the A-levels score prediction tool, online: <https://www.djib-xinwen.com/2020/08/le-royaume-uni-recule-dans-la-debacle-des-tests-de-niveau-a-liee-au-coronavirus/>.
12. Shoshana Zuboff, *The Age of Surveillance Capitalism: The Fight for a Human Future at the New Frontier of Power,* Profile Book Ltd (2019).
13. Elly A. Hogan and Viji Sathy, “8 Ways to Be More Inclusive in Your Zoom Teaching,” *Chronicle,* April 8, 2020, online: <https://www.chronicle.com/article/8-Ways-to-Be-More-Inclusive-in/248460>.
14. Anny Hakim and Natasha Singer, “New York Attorney General Looks Into Zoom's Privacy Practices,” *The New York Times,* March 30, 2020, online: <https://www.nytimes.com/2020/03/30/technology/new-york-attorney-general-zoom-privacy.html>; Lauren Feiner, “Zoom Strikes a Deal with NY AG Office, Closing the Inquiry Into Its Security Problems,” *CNBC,* May 7, 2020, online: <https://www.cnbc.com/2020/05/07/zoom-strikes-a-deal-with-ny-ag-office-closing-security-inquiry.html>; Valerie Strauss, “School Districts, Including New York City's, Start Banning Zoom Because of Online Security Issues,” *Washington Post,* April 4, 2020, online: <https://www.washingtonpost.com/education/2020/04/04/school-districts-including-new-york-citys-start-banning-zoom-because-online-security-issues>.
15. Kelly McCarthy, “The Global Impact of Coronavirus on Education,” *ABC News,* March 6, 2020, online: <https://abcnews.go.com/International/global-impact-coronavirus-education/story?id=69411738>
16. Simon Coghlan, Tim Miller, and Jeannie Paterson, 2020, *supra* note 9, p. 5.
17. Giacomo Panico, “University Cheating Might Be Up — But Don't Just Blame Students,” *CBC News,* July 19, 2020, online: <https://www.cbc.ca/news/canada/ottawa/university-cheating-might-be-up-but-don-t-just-blame-students-1.5618272>.
18. Ludwig Slusky, “Cybersecurity of Online Proctoring Systems,” *Journal of International Technology and Information Management,* 2020, 29-1, 56–86, p. 57, online: <https://www.semanticscholar.org/paper/Cybersecurity-of-Online-Proctoring-Systems-Slusky/80a084a512745c8b05c85a0870733d5cc435971c>.
19. Sheena Rossiter, “Cheating Becoming an Unexpected COVID-19 Side Effect for Universities,” *CBC News Edmonton,* June 21, 2020, online: <https://www.cbc.ca/news/canada/edmonton/cheating-becoming-an-unexpected-covid-19-side-effect-for-universities-1.5620442>.

20. Timothy H. Reisenwitz, “Examining the Necessity of Proctoring Online Exams,” *Journal of Higher Education Theory & Practice*, 2020, 20-1, 118, p. 123, online: <https://doi.org/10.33423/jhetp.v20i1.2782>.
21. *Ibid*; Neil Selwyn, Chris O’Neill, Gavin Smith, Mark Andrejevic, and Xin Gu, “A Necessary Evil? The Rise of Online Exam Proctoring in Australian Universities,” *Media International Australia incorporating Culture and Policy*, 2021, 1-16, pp. 1–2, online: https://journals.sagepub.com/doi/10.1177/1329878X211005862; Jake Evans, “ANU to Use Facial Detection Software on Student Computers in Response to Coronavirus Remote Exams,” *ABC News,* April 20, 2020, online: <https://www.abc.net.au/news/2020-04-20/coronavirus-anu-to-use-ai-spying-software-on-student-computers/12164324>.
22. Shawn Hubler, “Keeping Online Testing Honest? Or an Orwellian Overreach?,” *New York Times*, May 10, 2020, online:<https://www.nytimes.com/2020/05/10/us/online-testing-cheating-universities-coronavirus.html>.
23. Giamoco Panico, “U of O Students Wary of 'Extreme' Anti-Cheating Software,” *CBC News Ottawa*, July 2, 2020, online: <https://www.cbc.ca/news/canada/ottawa/exam-surveillance-software-university-ottawa-1.5633134>.
24. Radio-Canada, “L’utilisation d’un logiciel de télésurveillance pour les examens inquiète des étudiants,” *Radio-Canada,* août 18, 2020, online: <https://ici.radio-canada.ca/nouvelle/1727368/universites-examens-surveillance-covid-19-winnipeg-vie-privee>.
25. See <https://www.baneproctoring.com>.
26. Barbara Fedders, *The Constant and Expanding Classroom: Surveillance in K-12 Public Schools*, 97 N.C. L. Rev. 1673 (2019); Neil Selwyn et al., *supra* note 21, pp. 1–2.
27. Queen’s University, “Exams Services: Remote Proctoring,” *Queen’s University,* (n.d.), online: <https://www.queensu.ca/registrar/students/examinations/exams-office-services/remote-proctoring>.
28. Neil Selwyn et al., *supra* note 21, p. 2.
29. Ludwig Slusky, *supra* note 18, p. 57.
30. Neil Selwyn et al., *supra* note 21, p. 4.
31. Daniel Woldeab and Thomas Brothen, “Video Surveillance of Online Exam Proctoring: Exam Anxiety and Student Performance,” *International Journal of E-Learning & Distance Education*, 2021, 36-1, 1–26, p. 2, online: <https://www.proquest.com/docview/2572618998?parentSessionId=zl%2BfEpWBBJvqZTbySIdGx-UyimOQu0URzQYQ9uoPGi%2FY%3D&pq-origsite=primo&accountid=14701>.
32. David G. Balash, Dongkun Kim, Darika Shaibekova, Rahel A. Fainchtein, Micah Sherr, and Adam J. Aviv, “Examining the Examiners: Students’ Privacy and Security Perceptions of Online Proctoring Services,” 2020, p. 1, online: <https://arxiv.org/ftp/arxiv/papers/2106/2106.05917.pdf>.
33. Neil Selwyn et al., *supra* note 21, p. 12.
34. Université d’Ottawa, “Respondus LockDown Browser for Student,” *Université d’ Ottawa,* online: https://saea-tlss.uottawa.ca/en/teaching-technologies/academic-integrity-ouriginal-respondus/respondus.
35. Université d’Ottawa, “Welcome to the Virtual Campus powered by Brightspace (D2L),” *University of Ottawa*, online: https://uottawa.brightspace.com/d2l/home.
36. University of Manitoba, “Respondus,” *University of Manitoba*, (s.d.), online: <https://umanitoba.ca/centre-advancement-teaching-learning/technologies/respondus#respondus-lockdown-browser-and-respondus-monitor>.
37. Ryerson University, “Respondus LockDown Browser and Monitor virtual proctoring (integration with Brightspace),” *Ryerson University*, (n.d.), online: <https://www.ryerson.ca/courses/instructors/tutorials/integrations/virtual-proctoring-with-respondus-monitor/>.
38. Respondus, “Respondus Monitor,” (n.d.), online: <https://web.respondus.com/he/monitor/>.
39. Laval University, “Surveillance d’examen en télésurveillance avec Proctorio,” *University of Laval,* online: <https://www.enseigner.ulaval.ca/ressources-pedagogiques/proctorio>.

40. University of Concordia, "Proctorio," *University Concordia,* online: <https://www.concordia.ca/ctl/digital-teaching/proctorio.html>.
41. Meazure Learning, "Privacy Policy," *Meazure Learning*, February 17, 2022, online: <https://www.meazurelearning.com/privacy-policy>.
42. University of Toronto, "ProctorU: Privacy and Security Information Notice for Students," *University of Toronto*, online: https://teaching.utoronto.ca/resources/proctoru-privacy-and-security-information-notice-for-students/.
43. University of Waterloo, "Online Proctoring FAQs," online: https://uwaterloo.ca/extended-learning/online-proctoring/online-proctoring-faqs.
44. Athabasca University, "Using ProctorU for Online Exams," *Athabasca University,* online: <https://www.athabascau.ca/support-services/exam-services-support/using-proctoru-for-online-exams.html>.
45. Thompson Rivers University, "ProctorU," *Thompson Rivers University*, online: <https://www.tru.ca/distance/faq/faq-proctoru.html>.
46. Memorial University, "Online Proctoring," (s.d.), online: <https://citl.mun.ca/TeachingSupport/TeachOnline/OnlineProctoring.php>.
47. Proctor Exam, "Legal & Privacy," (s.d.), online: <https://proctorexam.com/privacy-and-data-security/>.
48. Polytechnique Montréal, "Mention légale relative à la protection des renseignements personnels," *Polytechnique Montréal,* (s.d.), online: https://share.polymtl.ca/alfresco/service/api/path/content;cm:content/workspace/SpacesStore/Company%20Home/Sites/etudes-web/documentLibrary/examens-finaux/Mention_legale_Proctor_Exam_Accommodement_Finaux.pdf?a=true&guest=true.
49. Examity, "All About Examity's Proctors," online: <https://www.examity.com/all-about-examitys-proctors/>.
50. University of Toronto, (s.d.), *supra* note 42.
51. E-Campus Ontario, online: <https://www.ecampusontario.ca/fr/proctortrack/>.
52. Queens' University, online: <http://www.queensu.ca/registrar/students/examinations/exams-office-services/remote-proctoring>.
53. Andrew Lupton, "Western Students Alerted About Security Breach at Exam Monitor Proctortrack," *CBC News*, October 15, 2021, online: <https://www.cbc.ca/news/canada/london/western-students-alerted-about-security-breach-at-exam-monitor-proctortrack-1.5764354>.
54. Nicholas Frew, "Software Used by U of Regina to Prevent Exam Cheating has Security Breach, Temporarily Shuts Down," *CBC News*, October 17, 2021, online: <https://www.cbc.ca/news/canada/saskatchewan/university-regina-proctortrack-security-breach-1.5765850>.
55. Regina University, online: <https://www.620ckrm.com/2020/10/19/u-of-r-test-monitoring-software-the-victim-of-a-security-breach/>. Western University, online: <https://ici.radio-canada.ca/nouvelle/1741804/informatique-pirate-fuite-vie-privee?fromApp=appInfoIos&partageApp=appInfoiOS&accesVia=partage>; E-campus Ontario, online: <https://www.ecampusontario.ca/fr/13855/>.
56. Simon Coghlan, Tim Miller, and Jeannie Paterson, *supra* note 10, p. 1.
57. Neil Selwyn et al., *supra* note 21, pp. 1–2.
58. Respondus, *supra* note 38.
59. Neil Selwyn et al., *supra* note 21, p. 2.
60. *Ibid*, at pp. 2–3.
61. *Ibid.*
62. *Ibid.*
63. The University of British Columbia, "Software for Securing & Proctoring Exams," *The University of British Columbia*, online: <https://isit.arts.ubc.ca/software-for-securing-proctoring-exams/>.
64. Simon Coghlan, Tim Miller, and Jeannie Paterson, *supra* note 10, p. 4.
65. *Ibid*; Todd Feathers, "Proctorio Is Using Racist Algorithms to Detect Faces," *Vice,* August 4, 2021, online: <https://www.vice.com/en/article/g5gxg3/proctorio-is-using-racist-algorithms-to-detect-faces>; Michelle Clark, "Students of Color are

Getting Flagged to Their Teachers Because Testing Software Can't See Them," *The Verge,* April 8, 2021, online: <https://www.theverge.com/2021/4/8/22374386/proctorio-racial-bias-issues-opencv-facial-detection-schools-tests-remote-learning>.
66. Todd Feathers, 2021, *supra* note 62.
67. Michelle Clark, *supra* note 62.
68. *"Online proctoring systems can be grouped into three classes: discretional live- proctor systems, automated (AI-enabled) systems, and hybrid (automated with a live proctor) systems"* Ludwig, *supra* note 15, p. 71.
69. The University of British Columbia, "UBC Senate Motions on Use of Remote Proctoring Software," *The University of British Columbia,* March 18, 2021, online: <https://lthub.ubc.ca/2021/03/18/ubcv-senate-motion-proctoring/>.
70. The University of British Columbia, online: <https://privacymatters.ubc.ca/sites/privacymatters.ubc.ca/files/PIA>.
71. "Automated Monitoring Paired with Human Review and Validation [...] During a Review+ session, the platform detects and flags patterns of suspicious activity, but an incident report is not created unless a certified proctor reviews the situation and confirms that the activity is not allowed in your exam rules." Meazure Learning, "Review+," *Meazure Learning,* online: <https://www.meazurelearning.com/products/proctoru-online-proctoring/review-plus>.
72. "During a Record+ session, the platform detects and flags patterns of suspicious activity, but an incident report is not created unless a certified proctor reviews the situation and confirms that the activity is not allowed in your exam rules." Meazure Learning, online: <https://www.meazurelearning.com/products/proctoru-online-proctoring/record-plus>.
73. Jane Bailey, Jacquelyn Burkell, Priscilla Regan, and Valerie Steeves, "Children's Privacy is at Risk with Rapid Shifts to Online Schooling Under Coronavirus," *The Conversation,* April 21, 2020, online: <https://theconversation.com/childrens-privacy-is-at-risk-with-rapid-shifts-to-online-schooling-under-coronavirus-135787>.
74. *Privacy Act* (R.S.C., 1985, c. P-21).
75. Canada Post Corp. v. Canada (Minister of Public Works), 1995 CanLII 3574 (FCA), [1995] 2 FC 110.
76. ProctorU, "Privacy Policy," *ProctorU,* online: <https://www.proctoru.com/privacy-policy>.
77. A-2.1 - *Act respecting Access to documents held by public bodies and the Protection of personal information*, Quebec, 2022.
78. Respondus, *supra* note 36.
79. *Ibid.*
80. *Ibid.*
81. Proctor Exam, Legal & Privacy, (s.d.), online:<https://proctorexam.com/privacy-and-data-security/>.
82. *Ibid.*
83. Examity, "Privacy & Security," *Examity,* (s.d.), online: <https://www.examity.com/features/privacy-and-security/>.
84. University of Toronto, "How Does Online Proctoring Work?" *University of Toronto,* (s.d.), online: <https://help.learn.utoronto.ca/hc/en-us/articles/115004994514-How-does-online-proctoring-work->.
85. Examity, "Product & Services Privacy Policy," *Examity,* (n.d.), online: <https://on.examity.com/V5/privacy>.
86. Examity, "Product Privacy Policy Appendix B," *Examity,* (n.d.), online: <https://www.examity.com/product-privacy-policy-appendix-b/>.
87. ECJ, 16 July, 2020, aff. C-311/18, DPC c. Facebook et Schrems, online: https://eur-lex.europa.eu/legal-content/EN/TXT/?uri=CELEX:62018CJ0311.
88. British Columbia: Freedom of Information and Protection of Privacy Act (FIPPA), R.S.B.C. 1996, c. 165.
89. Nova Scotia: Personal Information International Disclosure Protection Act (PIIDPA), Chapter 3 of the Acts of 2006 amended 2010, c. 35, s. 43.

90. Alberta: Personal Information Protection Act (PIPA), S.A. 2003, c. P-6.5.
91. An Act to modernize legislative provisions as regards the protection of personal information, SQ 2021, c 25.
92. *Ibid.*
93. FAQ Proctorio for students, online: <https://www.enseigner.ulaval.ca/ressources-pedagogiques/proctorio>.
94. For instance see PIPEDA Case Summary #2005-313.
95. Neil Selwyn et al., *supra* note 18, pp. 12–13.
96. Bill C-27, *An Act to enact the Consumer Privacy Protection Act, the Personal Information and Data Protection Tribunal Act and the Artificial Intelligence and Data Act and to make consequential and related amendments to other Acts, Part 3 Artificial Intelligence and Data Act,* 1st Sess, 44th leg, Canada, 2022, online: <https://www.parl.ca/DocumentViewer/en/44-1/bill/C-27/first-reading>.
97. *Ibid, Part 3 Artificial Intelligence and Data Act.*
98. *Ibid*, section 2 of *Artificial Intelligence and Data Act.*
99. Proposal for a Regulation of the European Parliament and of the Council Laying down harmonized rules on Artificial Intelligence (Artificial Intelligence Act) and amending certain union legislative acts (COM(2021) 206 final), online: https://eur-lex.europa.eu/legal-content/EN/TXT/?uri=CELEX:52021PC0206 [*EU AI Act*].
100. Supra note 97, section 3(1) of *Artificial Intelligence and Data Act.*
101. Including supervised, unsupervised and reinforcement learning, using a wide variety of methods including deep learning.
102. Including knowledge representation, inductive (logic) programming, knowledge bases, inference and deductive engines, (symbolic) reasoning and expert systems.
103. Bayesian estimation, search and optimization methods.
104. Supra note 93, section 4 of the *Artificial Intelligence and Data Act.*
105. T. Scassa, "Canada's Proposed AI & Data Act - Purpose and Application," *Blog*, August 8, 2022, online: https://www.teresascassa.ca/index.php?option=com_k2&view=item&id=362:canadas-proposed-ai–data-act-purpose-and-application&Itemid=80.
106. *Supra* note 97, section 5 of the *Artificial Intelligence and Data Act.*
107. *Ibid*, section 5(2).
108. *Ibid*, section 7.
109. T. Scassa, "Canada's Proposed AI & Data Act - Purpose and Application," *Blog*, August 8, 2022, online: https://www.teresascassa.ca/index.php?option=com_k2&view=item&id=362:canadas-proposed-ai–data-act-purpose-and-application&Itemid=80.
110. *Supra* note 97, section 5(2) of the *Artificial Intelligence and Data Act.*
111. *Ibid*, section 6.
112. *Ibid*, section 8.
113. *Ibid*, section 9.
114. *Ibid*, section 10(1).
115. *Ibid,* section 10(2).
116. *Ibid*, section 11.
117. *Ibid*, section 12.
118. See Mardi Witzel, "A Few Questions about Canada's Artificial Intelligence and Data Act: The Most Curious Aspect of the Proposed Law Is Also the Most Foundational Thing about It: The Overarching Governance Arrangement," *CIGI*, online: < https://www.cigionline.org/articles/a-few-questions-about-canadas-artificial-intelligence-and-data-act/>.
119. *Ibid.*
120. *Ibid.*
121. Annex II contains a list of Union harmonization legislation.
122. Annex III covers eight application areas for AI systems: Biometric identification and categorization of natural persons; Management and operation of critical infrastructure; Education and vocational training; Employment, workers management, and access to self-employment; Access to and enjoyment of essential private services and public ser-

vices and benefits; Law enforcement; Migration, asylum, and border control management; Administration of justice and democratic processes.
123. Government of Canada, "Directive on Automated Decision-Making," online: <https://www.tbs-sct.canada.ca/pol/doc-fra.aspx?id=32592>.
124. See also T. Scassa, "Canada's Proposed AI & Data Act - Purpose and Application," *Blog*, August 8, 2022, online: https://www.teresascassa.ca/index.php?option=com_k2&view=item&id=362:canadas-proposed-ai–data-act-purpose-and-application&Itemid=80.
125. [*EU AI Act*] *Supra* note 100, section 13.
126. *Ibid.*
127. *Ibid*, section 13b(iv).
128. *Ibid*, section 15(1).
129. The following artificial intelligence practices shall be prohibited:

(a) the placing on the market, putting into service or use of an AI system that deploys subliminal techniques beyond a person's consciousness in order to materially distort a person's behaviour in a manner that causes or is likely to cause that person or another person physical or psychological harm;
(b) the placing on the market, putting into service or use of an AI system that exploits any of the vulnerabilities of a specific group of persons due to their age, physical or mental disability, in order to materially distort the behaviour of a person pertaining to that group in a manner that causes or is likely to cause that person or another person physical or psychological harm;
(c) the placing on the market, putting into service or use of AI systems by public authorities or on their behalf for the evaluation or classification of the trustworthiness of natural persons over a certain period of time based on their social behaviour or known or predicted personal or personality characteristics, with the social score leading to either or both of the following:

(i) detrimental or unfavourable treatment of certain natural persons or whole groups thereof in social contexts which are unrelated to the contexts in which the data was originally generated or collected;
(ii) detrimental or unfavourable treatment of certain natural persons or whole groups thereof that is unjustified or disproportionate to their social behaviour or its gravity;

(d) the use of "real-time" remote biometric identification systems in publicly accessible spaces for the purpose of law enforcement, unless and in as far as such use is strictly necessary for one of the following objectives:

(i) the targeted search for specific potential victims of crime, including missing children;
(ii) the prevention of a specific, substantial, and imminent threat to the life or physical safety of natural persons or of a terrorist attack;
(iii) the detection, localisation, identification, or prosecution of a perpetrator or suspect of a criminal offence referred to in Article 2(2) of Council Framework Decision 2002/584/JHA 62 and punishable in the Member State concerned by a custodial sentence or a detention order for a maximum period of at least three years, as determined by the law of that Member State.

6 Blockchain, AI, and Data Protection in Healthcare

A Comparative Analysis of Two Blockchain Data Marketplaces in Relation to Fair Data Processing and the 'Data Double-Spending' Problem

Deepansha Chhabra
Department of Electrical and Computer Engineering, Faculty of Applied Science, The University of British Columbia, Vancouver, Canada

Meng Kang
Department of Electrical and Computer Engineering, Faculty of Applied Science, The University of British Columbia, Vancouver, Canada

Victoria Lemieux
School of Information and Institute for Computing, Information and Cognitive Systems, The University of British Columbia, Vancouver, Canada

CONTENTS

6.1 Introduction 126
6.1.1 Blockchain, AI, and Data Protection 127
6.1.2 The Data Double-Spending Problem 127
6.1.3 Chapter Outline and Contributions 128

DOI: 10.1201/9781003227656-10

6.2 Overview of the Case Study Data Marketplaces 128
6.2.1 Case Study One: Self-Sovereign Data Marketplace 128
6.2.1.1 Introduction 128
6.2.2 Solution Overview 130
6.2.2.1 Preliminaries 130
6.2.2.2 Roles 130
6.2.2.3 System Architecture 130
6.2.2.4 Computational Processing of Data 133
6.2.3 Case Study Two: Ocean Protocol Data Marketplace 133
6.2.3.1 Introduction 133
6.2.3.2 Preliminaries 133
6.2.3.3 Roles 134
6.2.3.4 System Architecture 135
6.2.3.5 Fixed vs Automatic Pricing 136
6.2.3.6 Computational Processing of Data 136
6.3 Findings 137
6.3.1 Principle 1: The Individual Controls and has Custody of Their Own Data 138
6.3.2 Principle 2: The Individual Consents to the Sharing of and Specific Use Their Data 141
6.3.3 Principle 3: The Individual Is Able to Withdraw Their Consent to Access and Use of Their Data 141
6.3.4 Principle 4: The Individual Has Been Appropriately Compensated for the Sharing of Their Data 142
6.4 Discussion and Conclusion 142
Acknowledgments 144
References 144
Appendix 146

6.1 INTRODUCTION

Personalized AI-driven health empowers individuals to take control of their health and has the potential to drive positive health and social benefits [1–3]. However, its adoption faces many challenges. We live at a time when the protection of individuals' health data is a major concern, with almost daily news about data breaches and mounting consumer distrust of large, centralized platforms that aggregate data and use it for secondary purposes without individuals' knowledge or consent [4–6]. People have become wary of for-profit companies or governments having control over their data, and, as a result, calls for data protection regulation are growing [7, 8]. If left unaddressed, concerns about data protection will act as a barrier to healthcare researchers and providers being able to access the real-world health data they need for AI-powered health advances [9, 10]. If these challenges can be overcome, however, it is possible to move society from a culture of treating illness to one of maintaining individuals' health through AI-powered personalized healthcare.

6.1.1 Blockchain, AI, and Data Protection

Blockchains have the potential to address data protection concerns in several ways: Data stored in a blockchains cannot be changed or deleted without detection (i.e., data are said to be 'immutable') [11]; blockchains' decentralized architecture means that no central authority controls data access and storage [12]; and blockchains can enable individuals to retain possession of the private keys that control access to and use of their personal health data (sometimes referred to as 'Self-Sovereign' data management, and often associated with the management of identity data, from which the moniker 'Self-Sovereign Identity,' or SSI, is derived) [13].

Several recent projects that focus on the application of blockchain in healthcare seek to leverage these capabilities to address data protection challenges associated with use of real-world healthcare data. For example, Carlini et al. [14] present the 'Genesy' platform for a blockchain-based fair ecosystem of genomic data. The Genesy platform aims to make available the data of users of genomic services connecting their data to research centers, pharmaceutical companies, hospitals, and geneticists. The capabilities of blockchain technology are used to notarize data and make them available for use by legitimate actors while safeguarding the data from unauthorized use to promote an ecosystem, and a fair marketplace, for all types of biomedical data. In a similar fashion, the 'PharmaLedger' project [15], which comprises a consortium of pharmaceutical and technology sector experts, patients, and hospitals, aims to deliver an open source, blockchain-based platform to build an ecosystem that promotes trust, transparency, and immutability and ensures privacy by means of decentralized and individual data control. In another example, Lemieux et al. [16] developed an SSI blockchain solution for secure and privacy-preserving health data sharing to explore how users would respond to navigating the complexity of blockchain technology for consenting to the sharing of their health data.

6.1.2 The Data Double-Spending Problem

While the literature on blockchain's application in healthcare recognizes data protection as a key constraint and design consideration, and projects like those cited above are rising to the challenge of designing with the protection of users' data in mind, solutions do not often consider the issue of unauthorized secondary use of data. When secondary use of data is not consented to by individual data subjects, we refer to this as 'data double spending' [17, 18]. Data double spending contravenes one of the principles of fairness in the processing of personally identifiable data, that is that individuals should consent to the sharing of and specific use their data. Other principles include that the individual should control and have custody of their own data, be able to withdraw their consent to access and use of their data and be appropriately compensated for the sharing of their data [19].

Data double spending is a practice that has become quite common in data-driven business models, including those in the healthcare sector. It occurs when individuals give consent to access and use of their data for a specific *primary* purpose, such as to receive a genetic test [20] or for purposes of COVID-19 contact tracing [21], but later the entity (often a for-profit company or overreaching government) subsequently uses the data for an unrelated *secondary* purpose. The individual originator of the data often is not aware of this secondary usage, might not agree with it, has not consented to it, does not benefit from it, and might even be harmed by it. Although data double spending is not necessarily illegal, many would consider it unethical, and it could have a 'chilling' effect on individuals' willingness to share their health data for socially beneficial purposes. Even if individuals are, themselves, not concerned about data double spending, possibly because they are not fully aware of the practice or of the harms that could arise from it, government agencies in liberal democracies charged with protecting their citizens against online threats are showing concern [22].

6.1.3 Chapter Outline and Contributions

This chapter seeks to contribute to research on the issue of data double spending. The chapter first describes two data marketplaces designed and implemented for individuals to share their health data for purposes of AI-driven health research. Case study one describes a solution design and implementation that uses the Hyperledger Indy/Aries protocol, the 'Self-Sovereign' data marketplace. The chapter then presents a second case study, which is an Ethereum-based solution called 'Ocean Protocol' data marketplace. Based on a comparison of implementations of both marketplaces, this chapter then evaluates the strengths and weaknesses of each solution vis-a-vis the principles of fairness in the processing of personally identifiable data. The chapter concludes with a consideration of the extent to which each solution operates to achieve fair data processing and protect individuals from data double spending, with the aim of contributing to a clearer articulation of the issue of data double spending, an assessment of how well each solution addresses the issue, and possible directions for future research aimed at countering data double spending.

6.2 OVERVIEW OF THE CASE STUDY DATA MARKETPLACES

In this section, we present an overview of our two case study data marketplaces preliminary to an analysis of how well each marketplace works to meet principles of fair data processing and prevents data double spending.

6.2.1 Case Study One: Self-Sovereign Data Marketplace

6.2.1.1 Introduction

Blockchain-based decentralized identity has emerged as a new privacy-preserving approach to identity management. Decentralized identity management provides for individuals to have full control over the use of their own digital identity data. While current decentralized identity-based solution designs are almost exclusively used for identity or certification authentication, decentralized identity systems have the potential to enable individuals with control over a much wider range of personal data for use beyond identity authentication, a capability that can be leveraged to address the data double-spending problem.

To ensure that the blockchain contains a consistent set of transactions, and that they are ordered correctly, each blockchain has its own consensus mechanism. To determine how the nodes come to an agreement about the contents, ordering, and insertion of transactions in a blockchain, as well as to determine how any changes to what has been written to the ledger would be detectable, a consensus mechanism must be implemented. The Bitcoin protocol achieves transaction blocks using proof-of-work, which must be proportional to the network's total computing power, to ensure a smooth and competitive operation. However, for such a large network to synchronize, the maximum throughput is restricted to a few transactions per second. Correspondingly, the Byzantine Fault Tolerance (BFT) consensus used in permissioned blockchains arose from the need to protect distributed systems from the threat of 'Byzantine failure,' where individual nodes in the network may be delayed in receiving new information from other nodes or may be sent maliciously-constructed information from malicious nodes. Existing BFT protocols demonstrate thousands of transactions per second throughput with dozens of nodes. The Redundant Byzantine Fault Tolerance protocol was used as the consensus mechanism in the Hyperledger Indy blockchain project, which operates as a decentralized public key registry in the SSI stack, because of its resilience and fast recovery properties.

SSI credential-based solutions have different privacy guarantees than other recordkeeping methods. SSI systems tackle a difficulty with recordkeeping that affects other types of distributed ledgers (i.e., personal information leakage from recording transactions on ledger). When clinical trial consent transactions are maintained on a blockchain, these records frequently contain personally identifiable information or metadata that could lead to reidentification of an individual, putting compliance with privacy and data protection laws at risk. SSI systems that do not record peer transactions on a ledger are designed to be highly privacy-preserving and comply with data protection legislation (e.g., the EU's General Data Protection Regulation) since they do raise the prospect of having to delete personally identifiable data from a ledger.

On the other hand, how to consume data while being able to protect data privacy and security and prevent leakage of sensitive information is a major challenge in data science today. Confidential computing was born to solve this problem [23]. A privacy-enhancing cryptography-based technology in the field of confidential computing, known as Fully Homomorphic Encryption (FHE) [24], was introduced in 2009. FHE allows computations on encrypted data while preserving the features and the format of the plain text. Thus, sensitive data, such as genomic and health data, could be stored in the cloud in an encrypted form without losing the utility of the data.

This case study discusses a novel design and implementation of a decentralized and privacy-preserving usage control enforcement infrastructure aimed at overcoming the limitations of centralized data usage control wherein (1) formulation of the data usage policies (i.e., licensing) and subsequent data usage is under the control of the individual data subject, (2) the data subject receives compensation for the use of their data, and (3) secondary use of the data that has not been authorized by the data subject is prevented. The solution was first presented in Kang and Lemieux [18]. It integrates a novel individual-centered digital rights management model with the use of decentralized identifiers (DIDs) and verifiable credentials (VCs) to establish a self-sovereign-oriented distributed trust architecture. FHE capabilities achieve additional privacy protection and data usage control. The solution design allows for data holders (who we envision also to be the data subjects) to configure a license for their data, integrates a payment system, uses VCs to provision data consumers with data access, and uses FHE to protect the privacy of data and prevent unauthorized secondary use, aimed at achieving a balance between value acquisition and data protection in the process of data sharing.

6.2.2 Solution Overview

6.2.2.1 Preliminaries

As a foundation to understanding the solution design, we present some preliminary concepts and technical primitives as follows:

- *Decentralized identifiers (DIDs)*, an emerging W3C standard for decentralized public key infrastructure which establishes decentralized trust roots.
- *DIDComm protocol*, a transport-independent protocol that uses DIDs to form and communicate over a cryptographically secure connection.
- *World Wide Web Consortium (W3C) Verifiable Credentials*, standard for cryptographically verifiable digital credentials.

6.2.2.2 Roles

Five roles are incorporated into this particular solution design:

- *Data Issuer*, the issuer of individual's data credentials.
- *Individual*, a person who is the controller of their own data and provides it, or shares it, with a consumer. In traditional data protection frameworks, this person might also be referred to as the 'data subject.'
- *Distributor*, the issuer of license credentials to consumers.
- *Data Manager* stores individual's data securely and provides the access token of the data.
- *Consumer*, person or entity with limited use of individual data under authorized conditions.

In this decentralized solution architecture, there can be multiple instances of each role (i.e., multiple participating Issuers, Individuals, Distributors, Data Managers, and Consumers). The basic interactions are as follows: The data is issued to the individual by means of VCs, and the data is encrypted by AES and FHE. The individual shares data by presenting a proof to the data manager. The data manager stores encrypted data in the cloud, and then issues a storage credential with the access token of the data. The individual sets up the data license using the distributor. The consumer purchases the data license through the distributor. Only after verifying the license credential, the individual provides the consumer with the token used to access the data and the key used to decrypt the data (AES only; the data always remain fully homomorphically encrypted) using a verifiable credential. Figure 6.1 shows the overview of roles and process flows as a Business Process Model and Notation diagram. Note that in this diagram the individual is identified as the 'Client.'

6.2.2.3 System Architecture

For ease of presenting the design, the architecture is divided into two parts. First, the data sharing and storage architecture is presented. Secondly, the data licensing and consumption architecture is presented. Figure 6.2 shows the architecture of data sharing and storage. The data issuer and data manager use a web app that

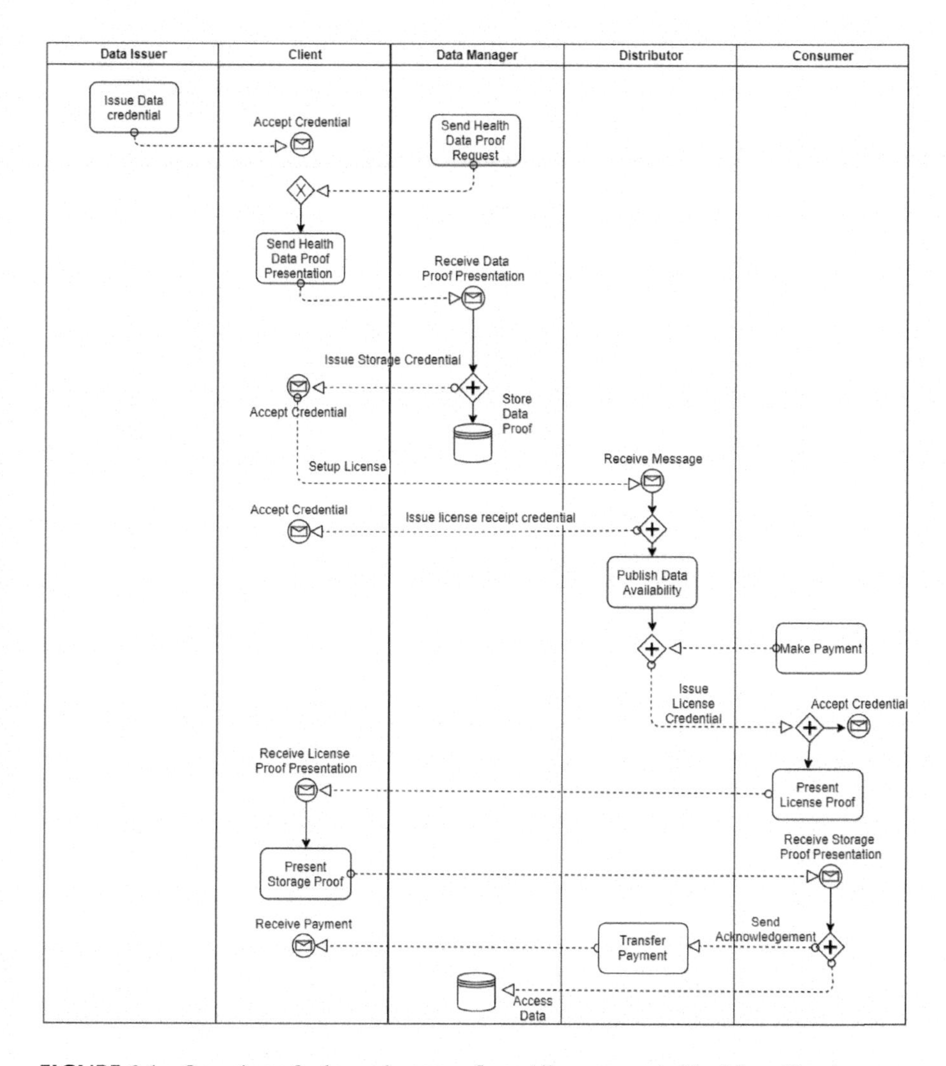

FIGURE 6.1 Overview of roles and process flows (diagram created by Meng Kang).

incorporates a Hyperledger cloud agent. The individual uses a mobile app with an agent. Agents can connect with other agents for messaging, credential issuing, and proof presentation.

To prevent the misuse of storage permissions, permissions have a limited duration of validity. Data never have a token that is valid indefinitely. This gives the individual, as controller of their own data, temporal control over active permission tokens as well as authorization capabilities. The signed URL is a query string authentication that has been used by most cloud storage providers as part of cloud storage access control mechanisms, which represents a concept of providing temporary access to specific resources [25]. All cloud service providers have an implementation of this technique, which is referred to as Signed URL in Google Cloud Storage [26], SAS in Microsoft

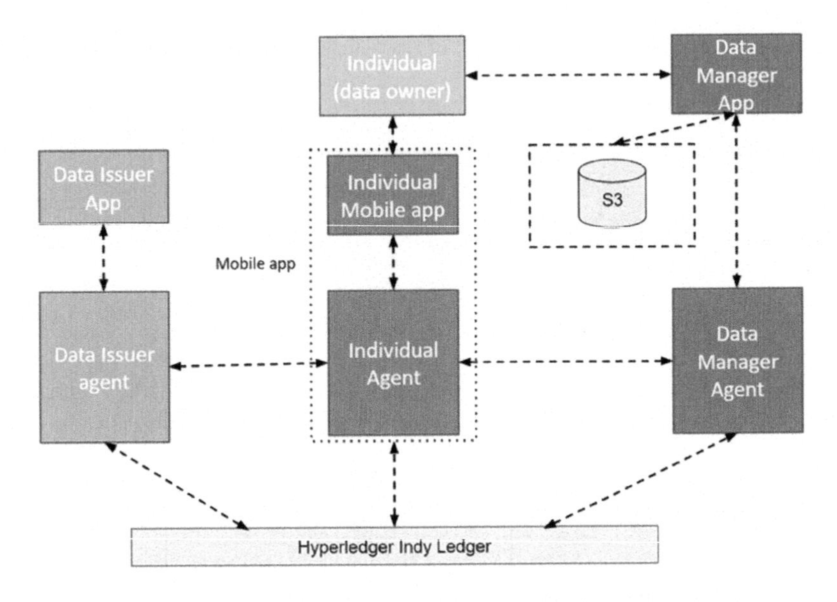

FIGURE 6.2 Data issuing and storage architecture (diagram created by Meng Kang).

Azure, and presigned URL in Amazon Web Services (AWS) [27] for purposes of granting temporary access. This solution uses the AWS presigned URL. This URL is embedded as an attribute of the storage credential.

Figure 6.3 shows the architecture of licensing and consumption. Similar to the previous part, the distributor and consumer have their web app with Hyperledger cloud agent. The Individual uses a mobile app with an agent.

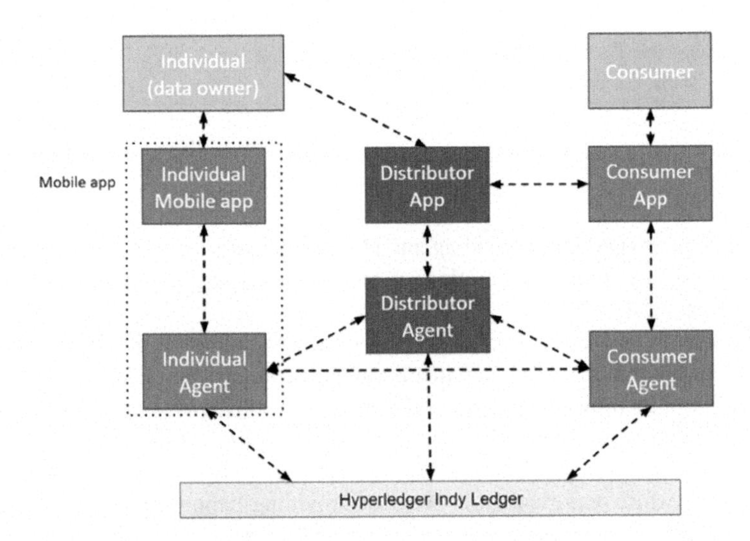

FIGURE 6.3 Licensing architecture (diagram created by Meng Kang).

6.2.2.4 Computational Processing of Data

The FHE adopted in this solution is a kind of cryptography scheme which allows operations on encrypted data without decryption. Currently, the Cheon-Kim-Kim-Song (CKKS) FHE scheme [28] is the most effective method of performing approximate homomorphic computations over real and complex numbers. By foregoing exact computations, the CKKS scheme achieves significant improvements in the ciphertext/plaintext ratio and algorithm speed. Therefore, this scheme is adopted in our solution as an encryption method for preserving the privacy of personal data. To prevent consumers from sharing homomorphically encrypted data, the design suggests that some level of protection and monitoring by the data manager as a service provider is required. These architectural enhancements require further research and involve the concept of a 'secure enclave.' Due to the effort and complexity of developing such an environment, this feature remains future work for this solution.

Snippets of code and screen shots from an implementation of the SSI data marketplace can be found in the Appendix.

6.2.3 Case Study Two: Ocean Protocol Data Marketplace

6.2.3.1 Introduction

Ocean Protocol (OP) is an Ethereum-based decentralized protocol which follows a systematic approach for unlocking data for AI analysis with the verification of eligibility, proofs of payments, and acceptance of terms and conditions of service level agreements [29, 30]. The overall goal of OP is to allow data providers to monetize their data in a privacy-preserving manner and consumers to obtain value from accessing and processing the data.

The data within the OP marketplace cannot be accessed directly if the compute-to-data approach is used. Data tokens that use the Ethereum ERC20 token standard are utilized to provide access to the data [30]. The data stays on individual data providers' premises and consumers can only perform computations on the data after accessing the data using a data token [30]. The OP decentralized database maintains a record of every transaction.

The data token can be transferred to a third party in exchange for the OP tokens [29, 30]. Any organization with access to a data token has compute access to the data. In other words, each data token provides a license to access the data [29, 30].

6.2.3.2 Preliminaries

As a foundation to understanding the solution design, we present some preliminary concepts and technical primitives based on V2.0-3.0 of OP as follows:

- *Data NFT,* a nonfungible ERC721 token representing copyright of a data service.
- *Data Token,* in the context of OP, an ERC20 token for gaining compute access to the data.
- *Ocean Token,* the token-based currency used in the OP marketplace.

- *Metadata, details used for asset discovery in the marketplace comprised of data asset, name*, creation date, etc. Each asset has a DID specific to it and a DID document for each metadata field. Fields and metadata can vary from one marketplace to the other.
- DID Document (DDO), a JSON document with metadata fields comprising of ID and signatures derived from the DIDs.
- Aquarius, a store of asset metadata. It is used for storing DID and DDO.
- Ocean DB manages the published data tokens in the OP marketplace.
- Operator Engine, a backend service responsible for providing necessary infrastructure for compute services and for executing the compute services in the Kubernetes Cluster (at the data provider's end).
- Automated Market Markers, used for the automatic price discovery in the marketplace. Example: Balancer, in which the data token sale is directly proportional to the price of data tokens.
- Gas, fee or pricing value which is required for the execution of transactions and contracts on Ethereum.
- KB Cluster, a platform at the data provider's end responsible for managing the compute service on data stored in containers.
- Compute-to-Data, a mechanism by which data consumers perform computations on the data on premises of the data provider.
- Compute Algorithm, the algorithm used in the compute-to-data service.
- EVM store, a secret store which hides the location of the data on the data provider's *premises.*

6.2.3.3 Roles

- *Data Provider,* in the context of OP, data providers are the organizations or group of individuals who are willing to sell their data in exchange for Ocean Token.
- *Brizo (also known as Provider),* serves as a data access control proxy of the data provider. It is the only component that can directly access the data location. It performs checks on chain for buyer permissions and payments; encrypts the data location URL and metadata during publication; decrypts the URL when the dataset is downloaded or a compute job is started; provides access to data assets by streaming data; and provides compute services (connects to C2D environment).
- *Operator Service,* service responsible for performing computations on the data based on the confirmation provided from the data provider's proxy server Brizo (Provider).
- *Data Consumer,* data consumer who is authorized to obtain compute access to the data for performing in exchange for Ocean token.
- *Keeper,* a node which is basically a set of smart contracts that maintains a registry of asset ownership and makes sure that all the nodes are running correctly within the system.

6.2.3.4 System Architecture

The OP marketplace has a common decentralized backend. However, there can be multiple frontends. The backend Architecture is presented below in Figure 6.4.

To interact in the marketplace, a data provider uploads their data to a storage location using AWS, Google Drive, or a decentralized data store such as Filecoin [29, 30]. In V2.0 of OP, the URL of the stored data is stored in a secret store – a cluster of EVMs [29], wherein each EVM stores a fragment of a data location URL – but this component has deprecated in a more recent version of the marketplace and the Brizo (Provider) now encrypts the URL and stores it on chain. The Brizo (Provider) also manages decryption when it receives a request for access to the data. To make the data available on the marketplace, the data provider 'publishes' the data using the Ocean Data NFT Factory, which creates a new data NFT. In an earlier version, a data NFT was not minted, and instead only ERC20 data tokens were used. However, since datasets are typically shared more than once or among many people, OP now mints a data NFT to define the ownership rights, with each NFT having one or more data tokens serving to separately define and grant access rights.

The Keeper node is used for running keeper contracts (a set of smart contracts) [29]. It makes sure that all the nodes are running correctly within the system. It is responsible for registering the data and maintains a registry of data ownership [29]. It makes use of the EVM and solidity technology and registers the metadata for the published data on chain using DIDs [29].

Data itself is not published on the OP marketplace; rather, it is the metadata and data tokens that are published [29, 30]. Metadata consists of fields such as the URL of the published data token, data token price, dataset name, description for the data, and so on [29]. DIDs for the data are hashed onto the metadata and are registered on chain using Keeper contracts [29].

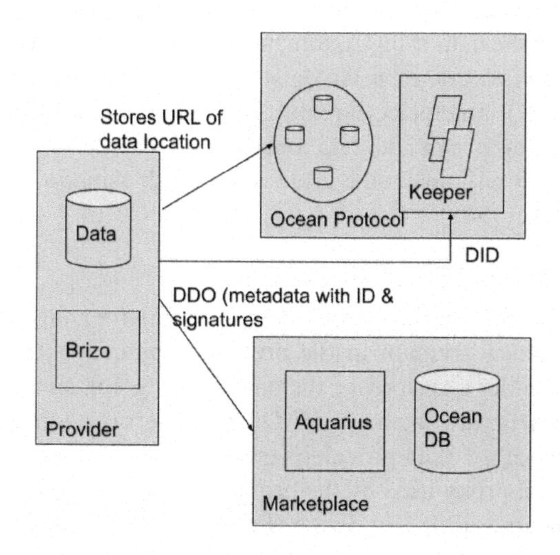

FIGURE 6.4 Data sharing and storage architecture (diagram created by Deepansha Chhabra).

Metadata has significant value as data consumers search for relevant datasets using metadata. Therefore, it is necessary to have concise and presentable metadata that gives consumers insight into the dataset [30]. Metadata along with the DID and DDO are stored in Aquarius, which is a metadata cache [29]. The marketplace consists of Aquarius and Ocean DB [29, 30].

After the data token for the data is published onto the marketplace, data consumers search for the relevant data using metadata keywords [29]. The Brizo (Provider) is an access control proxy used at the data provider's end [29]. It is responsible for providing access to the data consumers by means of a service URL. The data consumer requests the data, or compute access to the data, using the service URL.

The consumer is assigned the data token (access token). This access token is shown as a proof to the Brizo. The Brizo (Provider) receives the service request and provides the data, or compute access to the data, which is identified by its unique DID [29].

In the next step, automated verifications are used to verify the eligibility of the data consumer to access the data. For example, verification is carried out on whether the SLA has been signed or not, verification of payment, etc. [30]. The payments are received using an escrow account maintained by the Keeper [29]. The Brizo (Provider) verifies the paid transactions by communicating with the Keeper. After all the validations and verifications are completed, the Operator Engine is instructed to start computations over the data and return the results back to the consumer [29, 30]. Finally, the data provider can claim payment through an escrow account after the consumers are satisfied with the results of the computations [29]. This process is described in more detail below.

6.2.3.5 Fixed vs Automatic Pricing

Simple yet efficient smart contracts are used for setting a fixed pricing for the data (fixed OCEAN). However, there is another more flexible and better option to automate the pricing for the data using Automated Market Marker (AMM) [30].

Balancer is the most common Automated Market Marker which provides the advantage of adding marketplace liquidity through a single token [30]. Both data providers and consumers can add data tokens and Ocean tokens, respectively, into the AMM pool and it will detect the price and publish using initial data offering in a gas-efficient manner [30].

6.2.3.6 Computational Processing of Data

Compute-to-data is the approach to unlocking the data for analysis using AI, while allowing for the data to remain in the provider's premises, thus preserving the individual data provider's custody of the data and ongoing control [30]. The compute-to-data infrastructure includes the Operator service, Operator Engine, and Kubernetes Cluster at the data provider's end [30]. The Brizo (Provider), which is a data access control proxy used by the provider, receives a service access request from the data consumer [29]. The Brizo (Provider) verifies and validates the payments, acceptance of the SLA terms, and identity of the consumer and grants compute access to the data consumer along with the notification for publishing the compute algorithm [29].

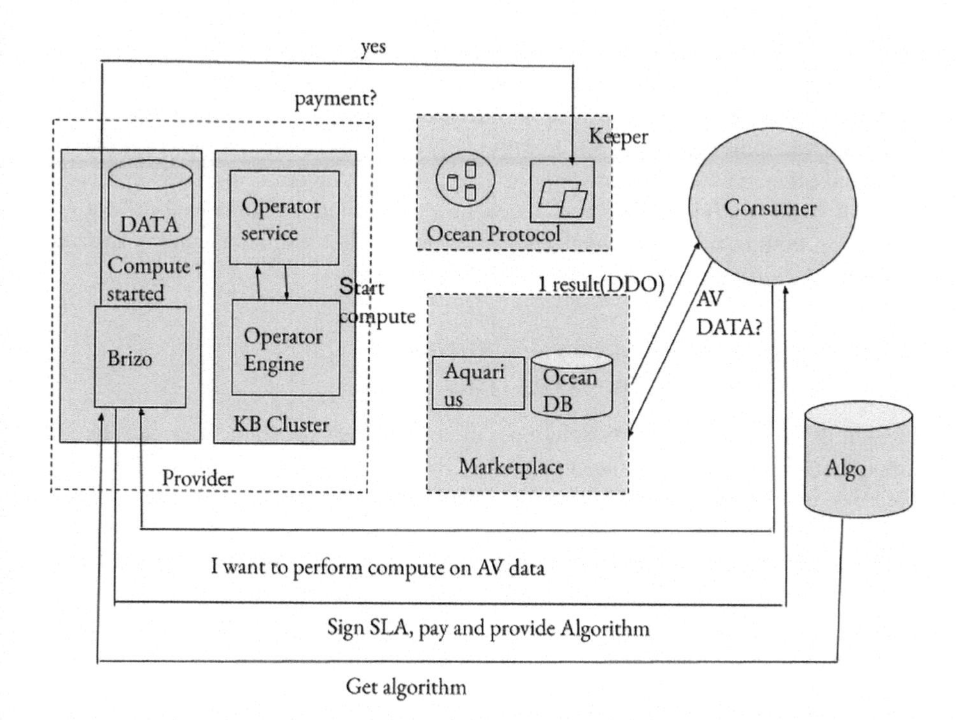

FIGURE 6.5 Compute-to-Data architecture.

As soon as the compute algorithm is published, a DID is assigned to the algorithm and the operator service is instructed to start the compute job on the published algorithm [29]. Final validations are performed by the data provider, who then claims the payment from an escrow account handled by the Keeper [29]. Finally, the Operator Engine starts the compute job by executing the Kubernetes Cluster, which performs computations on the data using the compute algorithm [29, 30].

Consumers are allowed to raise any queries about the status of their compute job [29, 30]. After the compute job is completed, the consumers are notified. Until their compute access expires, consumers can ask to restart the compute job if dissatisfied or if they wish to use a different algorithm [29]. After that, consumers must request access again. The execution logs and outputs are stored using AWS or Google drive storage [29, 30].

The following Figure 6.5 represents the compute-to-data architecture.

Snippets of code and screen shots from an implementation of the SSI data marketplace can be found in the Appendix.

6.3 FINDINGS

Having reviewed the technical primitives, system architecture, and data processing associated with each of the case study solutions, we now turn to evaluating how well each solution does in terms of fair data processing and solving the data double-spending problem. As previously mentioned, fairness in the processing of data generated and provided by individuals is supported when the individual controls and has custody of their own data; the individual consents to the sharing of and specific use their data; the individual is able to withdraw their consent to access and use of their data; the individual has been appropriately compensated for the sharing of their data; and others cannot subsequently use the data for secondary purposes not consented to by the individual. We now review how each solution addresses these design principles, both in respect to the primary sharing and use of the data and in respect to secondary sharing and use.

6.3.1 Principle 1: The Individual Controls and has Custody of Their Own Data

In each of our two case study solutions, the provider of the data, who may be an individual, is recognized as the controller of their own data, or 'owns' it. Moreover, both the solutions make use of DIDs and DDO. In the case of OP, DIDs are assigned as identifiers for datasets, whereas in the SSI solution, DIDs are used for actors, such as data providers, interacting in the marketplace. The use of dynamic DIDs in the case of the SSI solution helps ensure that the data cannot be traced back to a natural or legally identifiable person once shared. In this way, data consumers cannot discover the identity of individual data providers and the source of the data remains pseudonymous.

In the SSI solution, individual data providers are 'issued' data by data issuers from whom the data originate; for example, health data biomarkers might be generated in a lab from processing blood samples and then subsequently issued to the individual in the form of VCs. This model makes it possible to verify the authenticity of the data (i.e., its source of origin and its integrity), a capability that is facilitated by using Hyperledger Aries/Indy as a decentralized public key infrastructure; however, it does introduce reliance upon the data issuer for subsequent operations, since cryptographic proofs depend upon the data issuer's public key and data issuers may be able to revoke or render invalid VCs that they have issued. In the OP solution, on the other hand, the data issuer role does not exist. This is because, unlike the SSI solution which relies upon VCs to transmit and share health data, OP is an Ethereum-based solution that uses smart contracts and data tokens for providing access to third parties for provisioning access to or performing computation on the data. Smart contracts and the ability to tokenize data are not provided for by the Hyperledger Aries/Indy stack. OP operates as a permissionless ecosystem as of now but can be configured to be a permissioned ecosystem as well, whereas the SSI solution, because it relies on Hyperledger Indy, only operates as a permissioned ecosystem. In the OP solution, the 'keeper' maintains a registry of data ownership (synonymous with the data provider), but there is no mechanism by which the original source of the data, if other than the data provider, or its integrity can be cryptographically verified, unlike in the SSI solution, which supports such verification. This leaves the OP solution vulnerable to a problem commonly found in platforms offering artworks as nonfungible tokens (NFTs), being that it is not possible to ascertain whether the NFT represents ownership of an original or is merely a copied image of an original work.

In the case of the SSI solution using Hyperledger Indy/Aries, the individual does not permanently transfer the right to control the data, or data 'ownership'; rather, the solution provides the individual with a mechanism – a verifiable credential – for provisioning temporary access to a data consumer by presenting a proof based on the verifiable credential. In the OP solution, individuals can provision sharing of the data or the right to compute over the data to a consumer; however, in this solution, the mechanism used is a data token sent to the address of the consumer. When the consumer wishes to compute over the data, they send the data token to a wallet managed by the Brizo (Provider). Since access tokens could be sold and transferred to third parties, this increases the possibility of a data double spend in the OP solution.

Defining terms of use in the SSI solution is achieved by the distributor app, which allows individuals to specify service terms and issues license credentials to consumers. In the case of the OP solution, the data token and the Brizo (Provider) collectively fulfill the same role. Service terms are captured by the metadata of the data token while the Brizo (Provider) serves as a data access control proxy of the data provider. It is responsible for accepting data service requests, enforcement of service level agreements, verification of payments, and, finally, providing access to the data via a data token. In both solutions, components managing terms of use remain quite centralized, e.g., the individual relies upon a single distributor app, in the case of the SSI solution, and a single Brizo (Provider), in the case of the OP solution. The Brizo (Provider) component may be managed by the data provider, but most likely will be managed by the OP marketplace. Centralization of these service components does not provide the same defense against a single point of failure or interference (e.g., Sybil attacks) as would a more decentralized architecture. Given this, it is possible that an individual could lose control over their data through malicious tampering or failure of the component. Assuming normal operating conditions, however, individuals would have control of their data for purposes of any primary use.

In terms of data custody in both solutions, while data remains under the control of individual data providers until transferred under a license or service agreement, data cannot be said to be held by them, since each provider must rely upon, and trust, a third-party data storage provider, whether that be a centralized store, such as AWS or Google, or a decentralized store, such as Filecoin. Nevertheless, the individual data provider can choose which storage provider to trust, and the OP solution offers individuals the option to use a decentralized solution to avoid having to place trust in a centralized data storage provider. In a previous version of the architecture (V2), the URL for the stored data was kept in multiple secret stores (i.e., EVMs), with each EVM storing a fragment of a data location URL. None of the EVMs could decrypt the entire URL alone. This approach helped prevent unauthorized access to the data and reduced reliance on trusting the data storage provider, as only the originator of the data could access the data using their private key and the assembled fragments of the data storage URL. In a later version of the architecture, encryption/decryption of the URL is handled by the Brizo (Provider). When this component is operated by the OP marketplace, it requires the data provider to place their trust in the Provider to keep the location of the data secret and secure. In the SSI solution, unauthorized access to the data by an untrustworthy data storage provider is prevented by use of the combination of AES and FHE encryption. In this solution, the data provider holds the AES secret, which is only shared with the data consumer for purposes of gaining access to the FHE data once their claim to have paid for a license to access the data is verified. One could argue that the SSI solution is slightly superior in that the AES secret is generated and held by the data provider, whereas in the OP solution, there is a reliance on a third party. In addition, in the SSI solution, the data storage provider only ever stores fully homomorphically encrypted data, providing another layer of protection from an untrusted centralized data storage provider.

Regarding limiting unauthorized secondary use of data, i.e., preventing the data double-spending problem, both solutions remain quite limited in terms of the defense they provide. Each can provide for a time-limited access token that provisions access to the data for a specified period. In the case of the SSI solution, the data manager manages stored FHE data in the cloud and is responsible for issuing a storage credential with the access token for the data. Only after verifying the license credential, which the consumer purchases through the distributor app, does the individual provide a consumer with the token used to access the data and the AES key used to decrypt the data to gain access to the FHE data. The design also considers the fact that the data manager is likely to need to provide a computing environment to further restrict data use; thus, the presigned URL mentioned above should be more of an access token for the computing environment than just an access token for a data resource. Such an access token will provide access to a virtual environment which contains a computing resource (e.g., Amazon EC2) and the storage. However, at the present time, the solution does not provide this capability and the data is transferred into the custody of the consumer for processing. Nevertheless, it should be emphasized that, in the SSI solution, the data are never fully decrypted into clear text; they remain fully homomorphically encrypted thus limiting the possibility of unauthorized secondary usage; however, this approach does limit the type of computations that might be performed over the data.

In the case of the OP solution, the Brizo (Provider) provides access to the data by means of a service URL. Data consumers request direct or compute access to the data using the service URL. The Brizo (Provider) receives the request and provides access to data, identified by the data's unique DID. In a compute-to-data scenario, the consumer's algorithm is given access to the provider's data store to run computations over unencrypted data, in contrast to the SSI solution that provisions consumer access only to data that remains fully homomorphically encrypted.

Unauthorized secondary use of data could be achieved in either solution through techniques such as data mining to statistically reveal identities, so this also needs to be prevented to protect against a data double spend. Even if pseudonymous identities are used, if a single pseudonymous identity is reused many times, it may be possible to determine who the data provider is, and from there to discover transaction volume information and patterns of transactions that enable linking across multiple data shares.

In the OP ecosystem, the data token is independent of identity and does not rely on the use of identity credentials, so it does not present a path to finding out about the owner of data unless metadata embedded in the token reveals personally identifiable information. This is a possibility, however, and OP therefore provides for encryption of certain types of metadata for compliance with privacy laws.

Data tokens also involve the use of Ethereum public addresses, through which it could be possible to link to a natural or legal identity. The OP marketplace can also request the data provider to provide more information about their identity through, for example, tools like 3Box, which would allow for identification of a data provider. In addition, since the data to be shared or accessed are not encrypted, but analyzed in raw form, it is possible that aggregation of data could lead to a reidentification through statistical correlation.

In the SSI solution, the possibility of identity correlation is restricted by the use of dynamic decentralized identifiers that are not persistent, and thus difficult to correlate with a natural or legal identity. Further, data elements in the dataset to be shared that could be identifying (e.g., genomic data) are homomorphically encrypted and thus not provided in raw form. This also makes it difficult to aggregate data in ways that could lead to a future reidentification of the data provider.

6.3.2 Principle 2: The Individual Consents to the Sharing of and Specific Use Their Data

In each of our case study solutions, individuals consent to the sharing of their data by way of making it available to consumers via the data marketplace. In the SSI solution, users use the distributor app to define the specific terms (e.g., types of organizations, uses, and duration for which they are willing to make their data available for use). The distributor app manages these data usage conditions through requesting proofs from consumers based on VCs; for example, it might request a proof that the consumer is a hospital if the individual data provider has specified a desire only to share their data with this type of institution. In the OP solution, individual data providers can only specify service access variants, such as the duration of access, e.g., short, medium, or indefinite, which are then embedded into the data token as metadata. There is no manual license configuration option in OP at the present time. Access to the data token provides an automatic license. The Brizo (Provider) subsequently serves to enforce time-limited service terms in providing access to the data. Neither solution currently allows individual data providers to specify that the data cannot be used for any subsequent undefined secondary purposes, nor do these solutions provide technical means to prevent such unauthorized secondary usage, save by means of the mechanisms already identified above.

6.3.3 Principle 3: The Individual Is Able to Withdraw Their Consent to Access and Use of Their Data

Neither solution empowers an individual data provider to withdraw their consent to access and use of their data once a consumer has purchased a license to it. Moreover, once the consumer performs a computation using the data, such as training an algorithm or running an AI-based analysis process, it is impossible to remove the data from the ‘knowledge’ of the AI algorithm. However, at any time up to the point of providing the consumer with data access after purchase of access, the individual data provider may choose to withdraw their data from the marketplace. Further, the whole point of preventing a data double spend is to prevent consumers of data from overreaching when they have been provided with consent for data sharing and use, limiting the duration and purposes for which they can use the data to those consented to by the individual data provider and preventing the consumer from unilaterally, and without the consent of the data provider, sharing the data with another party for another purpose. As previously mentioned, in both solutions, controls to prevent this type of unauthorized secondary data use are still quite nascent.

6.3.4 Principle 4: The Individual Has Been Appropriately Compensated for the Sharing of Their Data

Both solutions support the operation of a data marketplace that provides a means by which individual data providers might be compensated for the sharing of their data. In the case of the SSI solution, it is the data provider who sets the price at which the data will be licensed for use. While this places a good amount of control in the hands of the individual data provider, it leaves them with the task of determining a fair market price, which, given market information asymmetries, might be difficult to do. In this regard, the OP solution offers better mechanisms for determining fair market price, such as its automatic pricing (AMM) service. Neither solution really grapples with the larger questions around the ethics of monetizing health data and the potential for socially harmful outcomes that such monetization might engender given unregulated pricing policies. For example, high prices for data could encourage individuals to sell access to their data even when such access might lead to reidentification and privacy breaches or when the trustworthiness of the consumer cannot be guaranteed. This is a matter that deserves the attention of, for example, bioethicists and health economists to assure ethical operational of both solutions.

In each solution, a token is used as compensation for data sharing and use. OP makes use of Ocean tokens in return for exchange of data tokens as well as for maintaining the OP marketplace. Hyperledger Indy/Aries has a TRC20 exchange token; however, the SSI solution uses the Tether USD stable coin, embedding a payment transaction hash as a variable in the license credential, which is subsequently presented to the individual data provider as a verifiable proof of payment before access is given.

6.4 DISCUSSION AND CONCLUSION

In this chapter, we have contributed to a definition and research on fair data processing and the issue of data double spending through analysis of two solutions designed for individuals to share their health data for purposes of AI-driven health research – one that uses the Hyperledger Indy/Aries protocol, the 'Self-Sovereign' (SSI) data marketplace, and the other an Ethereum-based solution, called OP data marketplace. Based on an implementation of data marketplaces using both protocols, we evaluated the strengths and weaknesses of each solution vis-a-vis principles of fair data processing in relation to data sharing and use and protecting individuals from data double spending. We found that both solutions are still quite limited in the protection they provide against unauthorized secondary use of data by a third party, even though they both align reasonably well with the main principles of fair data processing.

Neither solution really guarantees protection against the practice of unauthorized secondary use. This suggests that much more work could be done to extend the mechanisms to prevent data double spending – whether algorithmic or afforded by human governance – in each solution. For example, one novel research direction could be to explore how 'tokenomics,' or cryptocurrency-based incentive mechanisms, could be used in both solutions to disincentivize data double spending by punishing data double spenders (or rewarding consumers who use data appropriately).

Overall, we found that the SSI solution provided slightly greater protection against data double spending in that it does not use a persistent identifier for data providers, nor require them to provide identifying information or allow for data access rights (i.e., the access tokens) to be passed to a third party by the consumer, as in the case of the OP solution. The SSI solution also had an advantage in being able to specify certain conditions of data use in a license agreement, such as that the data consumer be of a certain type (e.g., for-profit or not-for-profit organization, from a certain jurisdiction, having a certain type of certification, etc.). Over and above this, the SSI solution also requires that consumers compute-to-FHE data, while the OP solution allows for computation over raw data. The requirement to compute over FHE data disincentivizes the copying and retention of data for some unspecified future secondary use. Using homomorphic encryption to safeguard data privacy, however, requires a significant amount of computational resources and is less efficient when compared to computation on unencrypted data. That is, in a commercialized scenario, data consumers may be forced to pay high computational resource prices in exchange for more stringent data protection. Thus, future research could be done to widen the scope of computations that can be performed over fully homomorphically encrypted data and increase the efficiency of processing in order to make the provision of FHE data more useful. Although processing of FHE data is computationally costly at present, the SSI solution does not rely on smart contracts, which also require the use of gas to operate. At time of writing, gas is quite expensive. This, in our view, would likely make the SSI solution less expensive to operate overall. Nevertheless, the OP solution offered superior mechanisms for price discovery in its data marketplace.

We also found that both solutions still rely upon many centralized trust components, so working toward a greater level of decentralization of the architectures and trust models for each solution would be beneficial to guard against both single points of failure and malicious manipulation that could lead to data double spending.

Each solution had its strengths and weaknesses in relation to the principles of fairness in data processing, specifically with respect to preventing unauthorized secondary data use, or data double spending, the focus of our research. We acknowledge that the principles we explored do not address wider issues of data privacy and security, nor do they offer an exhaustive set of design principles for fair data processing (indeed, [17] points to how very complex is the issue of interpreting what fairness means in the context of data protection legislation); nevertheless, our research helps refine understanding of the issue of data double spending and its relationship to the basics of fairness in data processing, which is an issue that has become of increasing concern to consumers and data protection authorities alike. Further, this chapter provides a comparative analysis of the capabilities of two very different blockchain-based data marketplaces in relation to the principles of fairness and prevention of data double spending. From this comparative analysis, it is possible to see how the principles of fairness in data processing design might be tackled in different ways by blockchain protocols and to identify avenues by which novel capabilities that prevent or reduce the likelihood of data double spending might be introduced into data marketplaces. We believe these types of innovations would prove to be socially valuable because, if individuals had greater ability to retain control over usage of their data, even after initially consenting to its use, it could generate more trust and a greater willingness to share real-world health data for use in preventing disease, identifying new therapies, and promoting health and wellness.

ACKNOWLEDGMENTS

The authors would like to acknowledge funding support for this research from the Natural Sciences and Engineering Research Council of Canada.

REFERENCES

1. Barrett, M., Boyne, J., Brandts, J., Rocca, B. L., De Maesschalck, L., De Wit, K., ... Zippel-Schultz, B. (2019). Artificial intelligence supported patient self-care in chronic heart failure: a paradigm shift from reactive to predictive, preventive, and personalised care. *EPMA Journal*, *10*(4), 445–464.
2. Rowe, J. P., & Lester, J. C. (2020). Artificial intelligence for personalized preventive adolescent healthcare. *Journal of Adolescent Health*, *67*(2), S52–S58.
3. Pataranutaporn, P., Danry, V., Leong, J., Punpongsanon, P., Novy, D., Maes, P., & Sra, M. (2021). AI-generated characters for supporting personalized learning and well-being. *Nature Machine Intelligence*, *3*(12), 1013–1022.
4. Jercich, K. (2021). The biggest healthcare data breaches of 2021. *Healthcare IT News*. https://www.healthcareitnews.com/news/biggest-healthcare-data-breaches-2021 (accessed: February 13, 2022).
5. Seh, A. H., Zarour, M., Alenezi, M., Sarkar, A. K., Agrawal, A., Kumar, R., & Ahmad Khan, R. (2020, June). Healthcare data breaches: insights and implications. *Healthcare*, *8*(2), 133.
6. Rosenbaum, E. (2018). 5 biggest risks of sharing your DNA with consumer genetic-testing companies. *CNBC*. https://www.cnbc.com/2018/06/16/5-biggest-risks-of-sharing-dna-with-consumer-genetic-testing-companies.html (accessed: February 13, 2022).
7. Gordon, W. J., Coravos, A. R., & Stern, A. D. (2021). Ushering in safe, effective, secure, and ethical medicine in the digital era. *NPJ Digital Medicine*, *4*(1), 1–3.
8. McGraw, D., & Mandl, K. D. (2021). Privacy protections to encourage use of health-relevant digital data in a learning health system. *NPJ Digital Medicine*, *4*(1), 1–11.
9. Mello, M. M., Adler-Milstein, J., Ding, K. L., & Savage, L. (2018). Legal barriers to the growth of health information exchange—boulders or pebbles?. *The Milbank Quarterly*, *96*(1), 110–143.
10. Xu, J., Glicksberg, B. S., Su, C., Walker, P., Bian, J., & Wang, F. (2021). Federated learning for healthcare informatics. *Journal of Healthcare Informatics Research*, *5*(1), 1–19.
11. Ben Fekih, R., & Lahami, M. (2020, June). Application of blockchain technology in healthcare: a comprehensive study. In *International Conference on Smart Homes and Health Telematics* (pp. 268–276). Springer, Cham.
12. Javed, I. T., Alharbi, F., Bellaj, B., Margaria, T., Crespi, N., & Qureshi, K. N. (2021, June). Health-ID: a blockchain-based decentralized identity management for remote healthcare. *Healthcare, 9*(6), 712.

13. Houtan, B., Hafid, A. S., & Makrakis, D. (2020). A survey on blockchain-based self-sovereign patient identity in healthcare. *IEEE Access*, *8*, 90478–90494.
14. Carlini, F., Carlini, R., Dalla Palma, S., Pareschi, R., & Zappone, F. (2020). The Genesy model for a blockchain-based fair ecosystem of genomic data. *Frontiers in Blockchain*, *3*, 57.
15. Ziegler, Y., Uli, V., & Wortmann, J. (2021). Blockchain innovation in pharmaceutical use cases: PharmaLedger and Mytigate. *Journal of Supply Chain Management, Logistics and Procurement*, *3*(4), 312–325.
16. Lemieux, V. L., Hofman, D., Hamouda, H., Batista, D., Kaur, R., Pan, W., ... Fraser, R. (2021). Having our "omic" cake and eating it too?: evaluating user response to using blockchain technology for private and secure health data management and sharing. *Frontiers in Blockchain*, *3*, 59.
17. Appari, A., & Johnson, M. E. (2010). Information security and privacy in healthcare: current state of research. *International Journal of Internet and Enterprise Management*, *6*(4), 279–314.
18. Kang, M., & Lemieux, V. (2021). A decentralized identity-based blockchain solution for privacy-preserving licensing of individual-controlled data to prevent unauthorized secondary data usage. *Ledger*, *6*.
19. Clifford, D., & Ausloos, J. (2018). Data protection and the role of fairness. *Yearbook of European Law*, *37*, 130–187.
20. Paul, K. (2021). Fears over DNA privacy as 23andMe plans to go public in deal with Richard Branson. *The Guardian*. https://www.theguardian.com/technology/2021/feb/09/23andme-dna-privacy-richard-branson-genetics (accessed: February 13, 2022).
21. Navarre, B. (2022). COVID-19 data-driven spark privacy and abuse fears.*US News*. https://www.usnews.com/news/best-countries/articles/2022-01-19/contact-tracing-biometrics-raise-privacy-concerns-amid-pandemic (accessed: February 13, 2022).
22. US National Counterintelligence and Security Center. (2021). China's collection of genomic and other healthcare data from America: risks to privacy and U.S. Economic and National Security. https://www.dni.gov/files/NCSC/documents/SafeguardingOurFuture/NCSC_China_Genomics_Fact_Sheet_2021revision20210203.pdf (accessed: February 13, 2022).
23. Sardar, M. U., & Fetzer, C. (2021). Confidential computing and related technologies: a review.
24. Gentry, C. (2009, May). Fully homomorphic encryption using ideal lattices. In *Proceedings of the Forty-First Annual ACM Symposium on Theory of Computing* (pp. 169–178).
25. Graupner, H., Torkura, K., Berger, P., Meinel, C., & Schnjakin, M. (2015, October). Secure access control for multi-cloud resources. In *2015 IEEE 40th Local Computer Networks Conference Workshops (LCN Workshops)* (pp. 722–729). IEEE, New York.
26. Sianipar, J., Willems, C., & Meinel, C. (2017, December). Signed URL for an isolated web server in a virtual laboratory. In *Proceedings of the 2017 9th International Conference on Education Technology and Computers* (pp. 218–222).
27. Boland, D. (2020). Securing Amazon Web Services (AWS) and Simple Storage Service (Amazon S3) Security. An article regarding Amazon Simple Storage Service security.
28. Cheon, J. H., Kim, A., Kim, M., & Song, Y. (2017, December). Homomorphic encryption for arithmetic of approximate numbers. In *International Conference on the Theory and Application of Cryptology and Information Security* (pp. 409–437). Springer, Cham.
29. Patel, M. (2020, December 14). Hitchhiker's guide to ocean protocol. *Medium*. https://blog.oceanprotocol.com/hitchhikers-guide-to-ocean-protocol-b238d3bbd450 (accessed: March 1, 2022).
30. Ocean Protocol Foundation & Big ChainDB. (2022, February 24). Ocean Protocol: Tools for the Web3 Economy. https://oceanprotocol.com/tech-whitepaper.pdf (accessed: March 2, 2022).
31. *Ocean Protocol*. (2021, November 9). GitHub. https://github.com/oceanprotocol (accessed: March 2, 2022).

APPENDIX

Here we present code snippets and screenshots showing the implementations of both case study solutions. The implementations are stored in a private GitHub repository, with access available on request.

1) SSI PROTOCOL

A) The data credential and storage credential

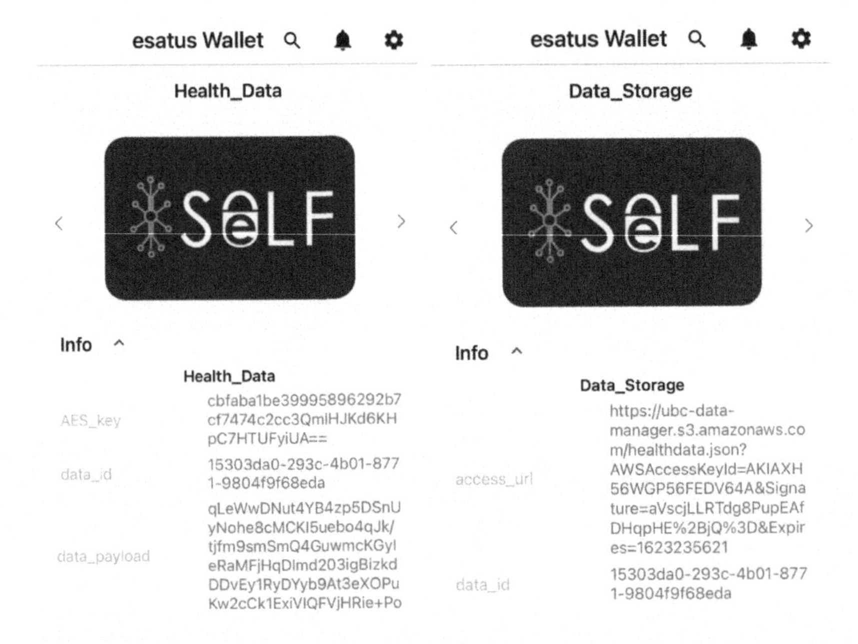

B) License Credential

C) License Proof


```
_id: ObjectId("60e0d8b32fb844d795b029f9")
presentation_exchang… : "c336e4db-5b2b-4554-a84b-d25bec44a8cc"
auto_present: false
connection_id: "aecb5668-dbc7-4f0f-9a25-21dcb7c2fe81"
created_at: "2021-06-29 21:37:54.970534Z"
initiator: "self"
› presentation_request: Object
› presentation_request… : Object
role: "verifier"
state: "verified"
thread_id: "2fcba6d0-a832-4379-9752-557183968690"
trace: false
updated_at: "2021-06-29 21:38:14.537352Z"
⌄ presentation: Object
  › proof: Object
  ⌄ requested_proof: Object
    › revealed_attrs: Object
    ⌄ revealed_attr_groups: Object
      ⌄ license: Object
          sub_proof_index: 0
        ⌄ values: Object
          ⌄ payment_method: Object
              raw: "Tether USD(trc-20)"
              encoded: "8697675681123806713592373599954969736276943614401754198699281475724755..."
          ⌄ data_available_dea… : Object
              raw: "2021-06-30 11:59:00 UTC"
              encoded: "7680940385823396195884050212325572251943764799984245284504846084634470..."
          › data_id: Object
          › distributor_address: Object
          › transaction_id: Object
          › payment_amount: Object
          › terms_and_conditio… : Object
    › self_attested_attrs: Object
    › unrevealed_attrs: Object
    › predicates: Object
  › identifiers: Array
verified: "true"
```


D) Storage Proof


```
_id: ObjectId("60e0077f135c0a80148a4d0c")
presentation_exchang… : "b105a0cd-1686-41e4-820d-d84a6b660536"
auto_present: false
connection_id: "4b3946f6-1a93-4562-b559-1e3ec29ad8ad"
created_at: "2021-06-29 06:45:19.130586Z"
initiator: "self"
› presentation_request: Object
› presentation_request… : Object
role: "verifier"
state: "verified"
thread_id: "304c3b75-9b76-411f-8bb6-86cff49ac032"
trace: false
updated_at: "2021-06-29 06:45:35.919271Z"
⌄ presentation: Object
  › proof: Object
  ⌄ requested_proof: Object
    ⌄ revealed_attrs: Object
    ⌄ revealed_attr_groups: Object
      ⌄ data_attrs: Object
          sub_proof_index: 0
        ⌄ values: Object
          ⌄ AES_key: Object
              raw: "cbfaba1be39995896292b7cf7474c2cc3QmlHJKd6KHpC7HTUFyiUA=="
              encoded: "7677873901358416902096447135361229517095963119793730132126485264390743..."
          ⌄ data_id: Object
              raw: "15303da0-293c-4b01-8771-9804f9f68eda"
              encoded: "987601128994125208221970906660343724061842917923963865784113563687219..."
      ⌄ storage_attrs: Object
          sub_proof_index: 1
        ⌄ values: Object
          ⌄ access_url: Object
              raw: "https://ubc-data-manager.s3.amazonaws.com/healthdata.json?AWSAccessKey..."
              encoded: "3441116483841489425444242711410071389187394892821724088690133947499022..."
          › data_id: Object
    ⌄ self_attested_attrs: Object
```

E) Data Sharing and Storage Algorithm

Algorithm 1: Data Sharing and Storage

1 *Individual* agent exchanges DIDs with the *Data Issuer* agent to establish a DIDComm connection channel.

2 *Data Issuer* uses the public key of the CKKS scheme (pk_{fhe}) to perform fully homomorphic encryption on the data of the individual. Then, *Data Issuer* uses the public key of the AES encryption scheme (pk_{aes}) to encrypt the data.

3 *Data Issuer* issues the data credential ($cred_{data}$) to the *Individual* agent, with the (pk_{aes}) as an attribute of the credential.

4 *Individual* accepts and stores the health credential in his/her mobile wallet.

5 *Individual* agent exchanges DIDs with the *Data Manager* agent to establish a DIDComm connection channel through the generated invitation QR code on the *Data Manager* app.

6 *Individual* agent sends the schema for proof request to *Data Manager* agent through a message.

7 *Data Manager* agent generates proof request (req_{data}) according to the schema sent by the Individual agent, and sends req_{data}.

8 *Individual* agent presents the health data proof $proof_{data}$ based on $cred_{data}$.

9 *Data Manager* agent verifies the $proof_{data}$ by checking the signature and DID of the *Data Issuer*.

10 **if** $proof_{data}$ *has been verified* **then**

11   *Data Manager* agent issues the initial storage credential ($cred_{storage_i}$) to the *Client* agent, with pre-signed URL and and pk_{aes} as attributes of $cred_{storage_i}$.

12 **end**

F) Data Licensing and Consumption Algorithm

Algorithm 2: Data Licensing and Consumption

1 *Individual* agent exchanges DIDs with the *Distributor* agent to establish a DIDComm connection channel.
2 *Individual* sets up the license schema on distributor app.
3 *Distributor* agent generates license receipt credential ($cred_{receipt}$), and send $cred_{receipt}$ to *Individual* agent.
4 **while** *Consumer selects the data with intention* **do**
5 *Consumer* agent exchanges DIDs with the *Distributor* agent to establish a DIDComm connection channel.
6 *Consumer* agent sends request for license credential ($cred_{license}$) to *Distributor* agent with the payment address.
7 **if** *Distributor receives the payment* **then**
8 *Distributor* adds the payment address to credential definition, issues $cred_{license}$ to *Consumer* agent.
9 *Individual* agent exchanges DIDs with the *Consumer* agent to establish a DIDComm connection channel by scanning QR code on distributor app.
10 *Individual* agent generates license proof request ($req_{license}$) and sends to *Consumer* agent.
11 *Consumer* agent presents license proof ($proof_{license}$) to *Individual* agent.
12 *Individual* agent verifies the $proof_{license}$ by checking the signature and DID of the *Distributor*.
13 **if** $proof_{license}$ *has been verified* **then**
14 *Individual* sends a request message for proof request of pre-signed URL ($req_{storage}$) to *Consumer* agent.
15 *Consumer* sends $req_{storage}$, which asks for URL from $cred_{storage_i}$, and pk_{aes} from $cred_{data}$.
16 **if** *pre-signed URL is expired* **then**
17 *Client* agent sends a message to *Data Manager* agent for a updated pre-signed URL credential ($cred_{storage_n}$).
18 *Data Manager* agent issues client agent $cred_{storage_n}$.
19 **end**
20 *Individual* agent presents the storage proof ($proof_{storage}$) to *Consumer* agent.
21 *Consumer* using pre-signed URL and pk_{aes} access health data. *Distributor* transfers the payment to the *Individual*
22 **end**
23 **end**
24 **end**

2) OCEAN PROTOCOL

A) Published Token

```
>>>
>>> DATA_ddo = ocean.assets.create(
...     metadata=DATA_metadata, # {"main" : {"type" : "dataset", ..}, ..}
...     publisher_wallet=alice_wallet,
...     services=[DATA_compute_service],
...     data_token_address=DATA_datatoken.address)

signing message with nonce 1: 0xbeD519EF79eE3b06b94751dFC8ce62587b8de6Cf, account=0x79d0A46eEe962550cA02dCC664C487Ea49f
INFO:ocean_lib.data_provider.data_service_provider:Asset urls encrypted successfully, encrypted urls str: {"encryptedDo
23c87a2e16a598725597de1e15cc62865c064abc9249d30e765ae85fa8a3732f9c811c6a03cea04d64ed764efa192518105e061bca40cd4438422cc
3d227eb746570814227e78eda2e30b5c87882add6ced710fb8254091c9e2b402a932fc891c5762086e881f0794a61a1dc267afe579e1e6186e0673b
525af3fb2698080abe277f68c88e80a9bb415b6d72bcaa1deaba663321be290404a997da341e85f0f16edb944bf8eed606e7bce882d673942e376a6
oint http://localhost:8030/api/v1/services/encrypt

INFO:ocean:Asset/ddo published on-chain successfully.
>>> print(f"DATA did = '{DATA_ddo.did}'")
DATA did = 'did:op:beD519EF79eE3b06b94751dFC8ce62587b8de6Cf'
>>>
>>> ALG_datatoken = ocean.create_data_token('ALG1', 'ALG1', alice_wallet, blob=ocean.config.metadata_cache_url)
>>> ALG_datatoken.mint(alice_wallet.address, to_wei(100), alice_wallet)
'0x4b5aee01dfa50247cab8f33f5675fc0d15a0c9d8222102ac81ac7cb7bc3fb8d0'
>>> print(f"ALG_datatoken.address = '{ALG_datatoken.address}'")
ALG_datatoken.address = '0xe16348c51a001059B2bA4Ad3c339442C9024d5A9'
>>> ALG_metadata =  {
...     "main": {
...         "type": "algorithm",
...         "algorithm": {
...             "language": "python",
...             "format": "docker-image",
...             "version": "0.1",
...             "container": {
...               "entrypoint": "python $ALGO",
...               "image": "oceanprotocol/algo_dockers",
...               "tag": "python-branin"
...             }
...         },
```

```
>>>
>>>
>>>
>>> did = asset.did  # did contains the datatoken address
>>> print(f"did = '{did}'")
did = 'did:op:4568CA67353c0db9eA07Fdf85Dc051468cF7397f'
>>>
>>> from ocean_lib.web3_internal.currency import to_wei
>>> data_token.mint(alice_wallet.address, to_wei(100), alice_wallet)
'0x9bfaf2bbdfcdb72b84de708ee832ba319f83f68047fbee111010f503fec0a410'
>>> from ocean_lib.models.btoken import BToken #BToken is ERC20
>>> OCEAN_token = BToken(ocean.web3, ocean.OCEAN_address)
>>> assert OCEAN_token.balanceOf(alice_wallet.address) > 0, "need OCEAN"
>>> pool = ocean.pool.create(
...     token_address,
...     data_token_amount=to_wei(100),
...     OCEAN_amount=to_wei(10),
...     from_wallet=alice_wallet
... )
BPool.newBPool(). Begin.
  pool_address = 0x1088701770DaA292dfacf04888D42A01C3bb054a
BFactory.newBPool(). Done.

>>> pool_address = pool.address
>>> print(f"pool_address = '{pool_address}'")
pool_address = '0x1088701770DaA292dfacf04888D42A01C3bb054a'
>>> from ocean_lib.common.agreements.service_types import ServiceTypes
>>> asset = ocean.assets.resolve(did)
>>> service1 = asset.get_service(ServiceTypes.ASSET_ACCESS)
>>> pool = ocean.pool.get(ocean.web3, pool_address)
>>> OCEAN_address = ocean.OCEAN_address
>>> price_in_OCEAN = ocean.pool.calcInGivenOut(
...      pool_address, OCEAN_address, token_address, token_out_amount=to_wei(1))
>>> from ocean_lib.web3_internal.currency import pretty_ether_and_wei
>>> print(f"Price of 1 {data_token.symbol()} is {pretty_ether_and_wei(price_in_OCEAN, 'OCEAN')}")
Price of 1 DT1 is 0.961 OCEAN (961117581407733563 wei)
```

B) Marketplace Displays

```
BFactory.newBPool(). Done.

>>> pool_address = pool.address
>>> print(f"pool_address = '{pool_address}'")
pool_address = '0x1088701770DaA292dfacf04888D42A01C3bb054a'
>>> from ocean_lib.common.agreements.service_types import ServiceTypes
>>> asset = ocean.assets.resolve(did)
>>> service1 = asset.get_service(ServiceTypes.ASSET_ACCESS)
>>> pool = ocean.pool.get(ocean.web3, pool_address)
>>> OCEAN_address = ocean.OCEAN_address
>>> price_in_OCEAN = ocean.pool.calcInGivenOut(
...     pool_address, OCEAN_address, token_address, token_out_amount=to_wei(1))
>>> from ocean_lib.web3_internal.currency import pretty_ether_and_wei
>>> print(f"Price of 1 {data_token.symbol()} is {pretty_ether_and_wei(price_in_OCEAN, 'OCEAN')}")
Price of 1 DT1 is 0.961 OCEAN (961117581407733563 wei)
>>> bob_private_key = os.getenv('TEST_PRIVATE_KEY2')
>>> bob_wallet = Wallet(ocean.web3, bob_private_key, config.block_confirmations, config.transaction_timeout)
>>> print(f"bob_wallet.address = '{bob_wallet.address}'")
bob_wallet.address = '0xBE5449a6A97aD46c8558A3356267Ee5D2731ab5e'
>>> assert ocean.web3.eth.get_balance(bob_wallet.address) > 0, "need ganache ETH"
>>> assert OCEAN_token.balanceOf(bob_wallet.address) > 0, "need ganache OCEAN"
>>> data_token = ocean.get_data_token(token_address)
>>> ocean.pool.buy_data_tokens(
...     pool_address,
...     amount=to_wei(1), # buy 1.0 datatoken
...     max_OCEAN_amount=to_wei(10), # pay up to 10.0 OCEAN
...     from_wallet=bob_wallet
... )
'0x8fd31833820d2771eef6089f6a0c77865f364f2dc80591330e4bd75adc9a54c7'
>>> from ocean_lib.web3_internal.currency import pretty_ether_and_wei
>>> print(f"Bob has {pretty_ether_and_wei(data_token.balanceOf(bob_wallet.address), data_token.symbol())}.")
Bob has 1 DT1 (1000000000000000000 wei).
>>>
>>> assert data_token.balanceOf(bob_wallet.address) >= to_wei(1), "Bob didn't get 1.0 datatokens"
>>>
```

C) Deployed Data Token

```
  type: 'aborted'
}
Alice account address: 6.95440762434068e+47
Deployed datatoken address: 0xD8ca63e000a680DE90aBc0957E0c28C7faFb997F
deepansha@deepansha-Inspiron-3542:~/ocean_js/barge/ocean-quickstart$ node index.js
Alice account address: 6.95440762434068e+47
Deployed datatoken address: 0x7e60F471f8be2b4e441A9D5022DEb3F45dF787e9
deepansha@deepansha-Inspiron-3542:~/ocean_js/barge/ocean-quickstart$
```

D) Data Token Shared

```
... )
>>> print(f"bob_wallet.address = '{bob_wallet.address}'")
bob_wallet.address = '0xBE5449a6A97aD46c8558A3356267Ee5D2731ab5e'
>>>
>>> # Alice shares access for both to Bob, as datatokens. Alternatively, Bob might have bought these in a market.
>>> DATA_datatoken.transfer(bob_wallet.address, to_wei(5), from_wallet=alice_wallet)
'0xeb7477ca9bdafc5c6f38a4464b9c79819f99ef92d4117e898e97a211d7aea446'
>>> ALG_datatoken.transfer(bob_wallet.address, to_wei(5), from_wallet=alice_wallet)
'0x2a06c798bb9d40bcbcd94212506eb171409599e7ff53cadc39de0f3687fa5d88'
>>> DATA_did = DATA_ddo.did  # for convenience
>>> ALG_did = ALG_ddo.did
>>> DATA_DDO = ocean.assets.resolve(DATA_did)  # make sure we operate on the updated and indexed metadata_cache_uri versi
>>> ALG_DDO = ocean.assets.resolve(ALG_did)
>>>
>>> compute_service = DATA_DDO.get_service('compute')
>>> algo_service = ALG_DDO.get_service('access')
>>>
>>> from ocean_lib.web3_internal.constants import ZERO_ADDRESS
>>> from ocean_lib.models.compute_input import ComputeInput
>>>
>>> # order & pay for dataset
>>> dataset_order_requirements = ocean.assets.order(
...     DATA_did, bob_wallet.address, service_type=compute_service.type
... )

INFO:ocean lib.data provider.data service provider:invoke the initialize endpoint with this url: http://localhost:8030/ap
```

E) Data Token Address

```
>> alice_wallet = Wallet(
..     ocean.web3,
..     os.getenv('TEST_PRIVATE_KEY1'),
..     config.block_confirmations,
..     config.transaction_timeout,
.. )
>> print(f"alice_wallet.address = '{alice_wallet.address}'")
lice_wallet.address = '0x79d0A46eEe962550cA02dCC664C487Ea49fC83C9'
>> from ocean_lib.web3_internal.currency import to_wei
>>
>> DATA_datatoken = ocean.create_data_token('DATA1', 'DATA1', alice_wallet, blob=ocean.config.metadata_cache
>> DATA_datatoken.mint(alice_wallet.address, to_wei(100), alice_wallet)
0xaf00f3c004b7a268b578dbc8007fdf81729298e6079c9df0553e27480f5d3995'
>> print(f"DATA_datatoken.address = '{DATA_datatoken.address}'")
ATA_datatoken.address = '0xbeD519EF79eE3b06b94751dFC8ce62587b8de6Cf'
>>
>> # Specify metadata & service attributes for Branin test dataset.
>> # It's specified using _local_ DDO metadata format; Aquarius will convert it to remote
>> # by removing `url` and adding `encryptedFiles` field.
>> DATA_metadata = {
..     "main": {
..         "type": "dataset",
..         "files": [
..       {
..         "url": "https://raw.githubusercontent.com/trentmc/branin/main/branin.arff",
..         "index": 0,
..         "contentType": "text/text"
..       }
..     ],
..     "name": "branin", "author": "Trent", "license": "CC0",
..     "dateCreated": "2019-12-28T10:55:11Z"
..     }
.. }
>> DATA_service_attributes = {
```

F) Compute-to-job started

```
...         algo_service.index,
...         ZERO_ADDRESS,
...         bob_wallet,
...         algo_order_requirements.computeAddress,
... )
>>>
>>> compute_inputs = [ComputeInput(DATA_did, DATA_order_tx_id, compute_service.index)]
>>> job_id = ocean.compute.start(
...     compute_inputs,
...     bob_wallet,
...     algorithm_did=ALG_did,
...     algorithm_tx_id=ALG_order_tx_id,
...     algorithm_data_token=ALG_datatoken.address
... )
INFO:ocean_lib.data_provider.data_service_provider:invoke start compute endpoint with this url: {'signature': '0x4f1abf6f3694ae09d378ce9ffefb1
130cded3def4520898fb3fa5240c3a694415cee5275c7c6e0f7850473d26ce8a46f8ccf9c026d0ea6d916231756d869cd7a1c', 'documentId': 'did:op:beD519EF79eE3b06
b94751dFC8ce62587b8de6Cf', 'consumerAddress': '0xBE5449a6A97aD46c8558A3356267EeSD2731ab5e', 'output': {'nodeUri': 'http://127.0.0.1:8545', 'br
izoUrl': 'http://localhost:8030', 'brizoAddress': '', 'metadata': {}, 'metadataUrl': 'http://localhost:5000', 'owner': '0xBE5449a6A97aD46c8558
A3356267Ee5D2731ab5e', 'publishOutput': 0, 'publishAlgorithmLog': 0, 'whitelist': []}, 'jobId': '', 'serviceId': 4, 'transferTxId': '0x59739a8
35837bba3ee0d4c475d80e18a41be8b73511cc8192a922c44b0af35dd', 'additionalInputs': [], 'userdata': None, 'algorithmDid': 'did:op:e16348c51a001059
B2bA4Ad3c339442C9024d5A9', 'algorithmDataToken': '0xe16348c51a001059B2bA4Ad3c339442C9024d5A9', 'algorithmTransferTxId': '0x03763eac140cffb5767
1afb2d860dc57f7cf54354b8e099500d021eb620cca9c'}

>>>
>>>
>>> print(f"Started compute job with id: {job_id}")
Started compute job with id: 054443bbc7614ec39bb2b499464d526d
>>>
>>>
>>>
>>>
>>>
>>>
>>>
>>>
```

G) Bob Downloads Data token

```
    return func(*args, **kwargs)
  File "/home/deepansha/ocean_js/barge/ocean-quickstart/market/test3/venv/lib/python3.8/site-packages/ocean_lib/ocean/ocean_assets.py", line 4
88, in pay_for_service
    raise InsufficientBalance(
ocean_lib.exceptions.InsufficientBalance: Your token balance 0 DT1 (0 wei) is not sufficient to execute the requested service. This service re
quires 1 DT1 (1000000000000000000 wei).
>>> print(f"order_tx_id = '{order_tx_id}'")
order_tx_id = '0x65e7db164c0b4ae78f8e3641c6b06e4bbdcf619822918c6b99b5abeec3f55351'
>>>
>>>
>>>
>>> file_path = ocean.assets.download(
...     asset.did,
...     service.index,
...     bob_wallet,
...     order_tx_id,
...     destination='./'
... )
exit()
signing message with nonce 0: did:op:4568CA67353c0db9eA07Fdf85Dc051468cF7397f, account=0xBE5449a6A97aD46c8558A3356267Ee5D2731ab5e
INFO:ocean_lib.data_provider.data_service_provider:invoke consume endpoint with this url: http://localhost:8030/api/v1/services/download?docum
entId=did%3Aop%3A4568CA67353c0db9eA07Fdf85Dc051468cF7397f&serviceId=3&serviceType=access&dataToken=0x4568CA67353c0db9eA07Fdf85Dc051468cF7397f&
transferTxId=0x65e7db164c0b4ae78f8e3641c6b06e4bbdcf619822918c6b99b5abeec3f55351&consumerAddress=0xBE5449a6A97aD46c8558A3356267Ee5D2731ab5e&sig
nature=0x5a52e808383abfb4248e3ea6220fed9201a6f8a7a1e013e2cf575ee885fce6764ca2cd768930a4365db54e1b182850243bebe6553824a63a2662bea54376c0541c&fi
leIndex=0
INFO:ocean_lib.data_provider.data_service_provider:Saved downloaded file in /home/deepansha/ocean_js/barge/ocean-quickstart/market/test3/dataf
ile.0x4568CA67353c0db9eA07Fdf85Dc051468cF7397f.3/branin.arff
```

H) Algorithm Published

```
signing message with nonce 2: 0xe16348c51a001059B2bA4Ad3c339442C9024d5A9, account=0x79d0A46eEe962550cA02dCC664C487Ea49fC83C9
INFO:ocean_lib.data_provider.data_service_provider:Asset urls encrypted successfully, encrypted urls str: {"encryptedDocument": "0x0455d32b04
8cc5bb0d24d3d9367ef8d56c022fe3d531cf683547a6ac13ac054f22ff0d1e5a0113b9697593a5421cfddd87b47b9ed76585c0b20517b44e857b03b2f9f7a42e893e944419652
fc02ce0d18a4f658b4d35a09ec092df857757079d4423f34058eecd1c0695f9fc7bcced083b009bf6b780c925457af01a3b96d85a94a50217a3a700385189a64ddfb03e9e2cd0
3ce6829849d620ec97a064a03fa528ee392eca6066ad4d1c924f9a2027b40febb20200aaa4bc77b723259a5dc968afe4c730304e9b950eee92a"}, encryptedEndpoint http
//localhost:8030/api/v1/services/encrypt
INFO:ocean:Asset/ddo published on-chain successfully.
>>> print(f"ALG did = '{ALG_ddo.did}'")
ALG did = 'did:op:e16348c51a001059B2bA4Ad3c339442C9024d5A9'
>>> from ocean_lib.assets.trusted_algorithms import add_publisher_trusted_algorithm
>>> add_publisher_trusted_algorithm(DATA_ddo, ALG_ddo.did, config.metadata_cache_uri)
[{'did': 'did:op:e16348c51a001059B2bA4Ad3c339442C9024d5A9', 'filesChecksum': 'e129c3020f75d45b0af5d49f4e51fbe88c1e4f77975e29fe67517e5332ab675
', 'containerSectionChecksum': 'b168c8f0a4da280cd05782167ac20a50ac9e34bd1ff80488421e50b480071684'}]
>>> ocean.assets.update(DATA_ddo, publisher_wallet=alice_wallet)
INFO:ocean:Asset/ddo updated on-chain successfully.
'0x531d01640280c5416b02ac25b0b7104bcd88f3acfd8c5d2211f3f8763c41e818'
>>> bob_wallet = Wallet(
...     ocean.web3,
...     os.getenv('TEST_PRIVATE_KEY2'),
...     config.block_confirmations,
...     config.transaction_timeout,
... )
>>> print(f"bob_wallet.address = '{bob_wallet.address}'")
```

I) Data Sharing and Storage Algorithm

The following algorithm shows the process of data sharing and storage.

ALGORITHM: 1 Data Sharing and Storage

---

1. **Create** a Meta mask Wallet to interact with the Ethereum Blockchain.
2. **Download** Browser extension & pin to browser for easy access.
3. **Remember** (and keep secret) private key phrase.
4. **Go** to Ocean protocol Market website
5. **Connect** to the Meta mask Wallet
6. **Go** to the publish page option
7. **Fill** in the Publish Form.
8. **After** clicking submit, approve transactions in Wallet (Here you can see Meta mask wallet)
9. **Deploy** a new data token contract
10. **Interact** with the Contract
11. **Find** the publishing dataset option
12. **Create** Pricing (Fixed or Dynamic)
13. **If** Fixed Pricing.

    Go to the published Asset.

    If Pricing is not set

    Click create pricing Button (set value of data token with respect to ocean token

    else

    Fixed Pricing already set

    else

    Let the market discover the right price for data derived from Defi with liquidity pools. It is implemented using Balancer, an AMM protocol as Liquidity pool holding Ocean tokens & data tokens.

    Data token has been successfully published on Ocean Market.

J) Compute-to-Data Algorithm

The following algorithm explains the steps involved in compute to data.

ALGORITHM: Compute-to-Data Algorithm

---

1. **Set** up computing architecture (Aquarius, Operator service & operator Engine)
2. **Publish** data asset on Ocean & receive DID for published data asset
3. **Receive** a compute access request to ocean and for publishing the algorithm to be used to compute on providers' data by the consumer.
4. **Sell** access to train AI model on the data via compute service to consumers.
5. **Notify** consumer to publish the algorithm on Ocean and receive DID (algo DID) for algorithm
6. **Notify** consumer to sign the SLA and pay for the compute service via escrow smart contract account which will be used as payment for compute service
7. **Receive** compute service request from consumer and operator service starts compute.
8. **Find datasets** and algorithms in Aquarius (data DID and algorithm DID) (fixed identifier not dynamic)
9. **Instruct** the operator engine to initiate compute job.
10. **Start** compute task After validations of inputs (request is valid and signed SLA).
11. **Spin** a Kubernetes cluster process within the given parameters of data and algorithm.
12. **Respond** to data consumers if they enquire about status of compute job.
13. **Restart** compute execution if customer isn't satisfied, with the same or a different algorithm, until the compute access expires.
14. Inform the data consumer after completion of compute job.
15. **Share** the URL of storage of execution logs and outputs generated,

7 Cyber Influence Stakes

Rachel Ladouceur and Fehmi Jaafar
Department of Mathematics and Computer Science,
Québec University at Chicoutimi, Québec, Canada

CONTENTS

7.1 Introduction 156
7.2 Background 156
7.2.1 Context 156
7.2.2 Types of Influencers 157
7.2.2.1 Nation-states 157
7.2.2.2 Hacktivists 158
7.2.2.3 Political Actors 158
7.3 Strategies Under the Cyber Influence 159
7.3.1 Military Objective 159
7.3.2 Political Objective 159
7.3.3 Economical Objective 160
7.3.4 Ideological Objective 160
7.4 Cyber Influence on Social Media 161
7.4.1 Reasons for Its Popularity 161
7.4.2 How is Disinformation Spreading so Fast and Wildly on Social media? 162
7.5 Techniques of Cyber Influence 162
7.5.1 Trolls and Bots 162
7.5.2 Fake Accounts 163
7.5.3 Hijacking Existing Social accounts/hashtags 163
7.5.4 Deepfakes 164
7.6 Countermeasures of Cyber Influence 164
7.6.1 Countermeasures Used by Social media 164
7.6.1.1 Redirection 165
7.6.1.2 Labeling 165
7.6.1.3 Fact-checking 165
7.6.2 Detection of Trolls and Bots 166
7.6.3 Deepfakes' Authentication and Detection 168
7.7 Discussion 169
7.7.1 Improving Countermeasures 169
7.7.2 Cyber Influence Impacts 169
7.8 Conclusions and Future Work 170
References 171

DOI: 10.1201/9781003227656-11

7.1 INTRODUCTION

Cyber influence is fake news propaganda on social media. Threat actors are doing this for a specific reason, which can be political, economic, or military purposes. Does anyone have read fake news or watched fake videos and fake photos about Russo-Ukrainian? Yes. The cyber threat actors want to gain the public hearts; they send fake news to anyone, mainly on social media. That's the new part and the black side of social media.

Social media plays a big part in spreading fake news because it provides practical ways that make disinformation activities easy to do, fast, and cheap. Instead of humans, robots, called bots, send automated messages to thousands of users. Indeed, threat actors use different techniques to make fake news viral. Those well-known platforms, such as Facebook and Twitter, are the most used to communicate up to now. Studies have been realized on them to understand influence comportment. But others like YouTube and TikTok are social media that no studies have been realized as of today. There are real black boxes.

Countermeasures can be applied by users and by social media themselves. In this chapter, we will show that it's impossible to eradicate fake news because measures are not enough automated and they are not proactive or, in other words, preventive. The consequences are that cyber influence has negative impacts on society, from anti-vax conspiration which has altered world immunity to political interference. Indeed, General Paul Nakasone, director of the United States National Security Agency and head of Cyber Command declared that "foreign influence operations would be the next big disruptor" (de Rochegonde & Tenenbaum, 2021).

Who are influencers? We do not think we know exactly who they are. Social media platforms don't have the responsibility to track them. Indeed, this is one of the most fundamental problems with social media. Some influencers are megadonors, who have the power, in other words, who have the control: as an example, the Mercer family, who had played a big role in the 2020 US presidential election and financed last year over nearly $20 million in secret contributions. As another example, we can cite Steve Bannon, who was Trump's advisor and, at the same time, was sitting on the board of Cambridge Analytica. In addition, those companies have created fake social profiles to influence American electors to vote for Trump.

7.2 BACKGROUND

7.2.1 Context

In this section, we will define what is cyber influence and then we will portray the type of influencers.

We notice an absence of standardized definitions of cyber influence. Here below are the frequency terms to describe cyber influence, by order of importance: fake news, misinformation, propaganda, disinformation, influence campaigns/operations, and information operations/war (Smith & Thompson, 2020).

Those words are all about FALSE information. False information can be intentional (disinformation) or not (misinformation). Disinformation is "information that is false and deliberately created to harm a person, social group, organization or country social platforms are conduits of the information disorder. Misinformation is generally used to refer to misleading information created or disseminated without manipulative or malicious intent" (Berge, 2018). Propaganda is the promotion of ideas.

Cyber influence is done to obtain an advantage over his adversary via Internet (Grimes, 1978). We will see further that it could be a political, economic, or ideological advantage. So cyber influence is generally a part of a whole strategy but, to avoid a military response, the cyber threat actors adapt their digital techniques to stay anonymous and must act below the threshold of war (Henschke et al., 2020).

In general, fake news, propaganda, and disinformation are all synonyms. Cyber influence is spreading fake news or disinformation on social media to promote a point of view or political cause and to have a huge influence without spending a lot of money. It's worth it.

It's important to distinguish the word "influence" and "interference." Interference is activities that violate the law (Government of Canada, 2021). But the line between legal and illegal is sometimes fuzzy.

7.2.2 Types of Influencers

The Communications Security Establishment of Canada has identified cyber threat actors' profiles. Those who can influence the democratic process are mainly those actors: nation-states, hacktivists, and political actors (Government of Canada, 2018). We will detail each of these actors.

7.2.2.1 Nation-states

Influenced countries, such as the US, China, and Russia, have the most sophisticated means and use cyber influence for economic, ideological, and political purposes. As we have seen during the 2016 US presidential election, threat actors have tried to amplify social disputes, like Black Lives Matter. As noted in The Canadian cyber threat assessment, "they try to emphasize existing friction in democratic societies around political, economic issues, and other values such as human rights and liberty. They adapt their dialogues to the new context and, change their strategies" (Government of Canada, 2020). They even can pay journalists to write biased articles to give more credibility to the messages; we have seen that in the past.

According to this study (Martin & Shapiro, 2019), Russia has been the leading country to use cyber influence strategies around the world since 2017. Indeed, the Russian army, the Kremlin, largely finances these activities. Their goal is to erode the liberal model, by exploiting the internal division in liberal political parties in Europe and US democracies and amplifying social division.

Even if the US is the most targeted country with 38% of cyber influence case studies since 2013 (Martin & Shapiro, 2019), this state used as well cyber influence to gain economic and ideological purposes. As for examples noted in de Rochegonde and Tenenbaum (2021), cyber influence can be used to ban extremist ideologies such as Islamist organizations. It could also be used to disrupt an adversary's activities as was the case, in 2016, when the US wanted to confuse and influence the perceptions of Islamic State of Iraq fighters.

7.2.2.2 Hacktivists

Hacktivists are hackers' activists with a political or ideological goal. For example, a well-known group is Anonymous. They are known for their various cyberattacks against several governments around the world.

When the war started in Ukraine in 2022, several other groups (e.g., Squad 303) joined Anonymous to declare war on Russia. A multitude of cyberattacks, such as denial of service (DDoS), made servers unfunctional: Kremlin website and television channel RT website; malware wiper, which have stolen information from the servers of Roskomnadzor, the Russian media policeman. Those malware took control of several Russian news channels, to show images of Russian attacks (Maria, 2022).

Russia has its own cyberattack agency called IRA, the Internet Research Agency. Since 2014, several cyberattacks have occurred in Ukraine, such as electoral inference to government and public services blackout (e.g., gas, healthcare, etc.). It was caused by destructible malware, KillDisk, and Industroyer, and each of them has caused an electricity blackout (de Rochegonde & Tenenbaum, 2021).

7.2.2.3 Political Actors

Political actors are motivated by winning their elections. And they can be well funded by rich people who have their own interests and have the power to manipulate the election. This is the case of the Cambridge Analytica and Aggregate IQ scandal. At the time, those companies were owned by the billionaire Robert Mercer and headed by Donald Trump's key adviser, Steve Bannon. In 2014, Facebook accounts were sold to those companies. In 2016, for Trump's US presidential election purpose, those companies have created profiles of 230 million Americans and then targeted them with political advertisements, as reported by Derosa (2018).

Furthermore, those companies provided also services to British political organizations, which were involved in the Leave the European Union campaign during the 2016 Brexit referendum (Cadwalldr, 2018).

Robert Mercer has purely political interests. He has invested millions of dollars to manipulate elections such as Trump's presidential election in 2016, and to influence groups who are against climate change (Delevingne, 2014). We think that those influencers or megadonors are not enough followed for their underground activities on social media. They are not labeled because big influencers are usually unknown.

Other influencers have been used during the 2016 presidential election, such as the QAnon movement, who have sent threatening messages to voters (Laviola, 2018).

Finally, Donald Trump is a political actor who retains great cyber influence. Even though he is no longer the president of the US, he continues to influence his supporters. Banned from social networks such as Twitter, he decided to launch his own social network. According to Lachapelle (2022), Truth Social will be launched and expected to have no less than 75 million users worldwide, while Facebook's users in its first year were around one million. It could be a way for Trump to manipulate information to influence those undecided voters for the next election. In May 2022, it was the most downloaded platform on Apple Store.

In sum, political actors and those who financed them could be those who influence the most. We don't know them exactly and what influenced groups they are invested in and how they manipulated information and people on social media. But the first question to ask is how can we define "influencers"? Are those who can change the behaviors or opinions of people?

7.3 STRATEGIES UNDER THE CYBER INFLUENCE

All strategies cited below can be merged into a global strategy against the adversary.

7.3.1 Military Objective

It has been demonstrated that cyber influence effectively achieves military objectives (MacKenzie, 2018). Important countries use cyber influence to sow division, confusion, and damage the credibility of the adversary. Taking advantage of the adversary is basically winning a part of the battle, it is to have one more advantage than him.

The new part is that social media serves as communication to give the latest news about the war where each party uses it to promote military action to the public. A very recent case where cyber influence has been used in a military context is the Russo-Ukrainian war in 2022. Russia has opted for a hybrid strategy: military intervention, cyber influence, and cyberattacks. Furthermore, this country restricted information to reinforce its message to have a positive perception of its action and strengthen the commitment of its people to fight for its legitimate cause.

According to Le Cointre, the Chief of Staff of the Armed Forces of France, "The objective of information war will be to fight against attempts to destabilize information on our space" (de Rochegonde & Tenenbaum, 2021).

7.3.2 Political Objective

Cyber influence in the political context aims at harming democratic institutions, such as undermining the candidates, sowing doubt about the legitimacy of the electoral process and the results, intimidating voters to not vote, increasing social division, etc.

Cyber influence activities generally increase around elections (Government of Canada, 2020). Fake news was much more numerous in the 2016 US election period than the usual period. In fact, the French fake news law 2018 establishes that three months before election day, no fake news will be tolerated, and actors could be eligible for possible legal action.

Russia has used cyber influence to manipulate voters' opinions occurred in the recent years: the US 2016 presidential election, the US 2018 midterms, and the 2016 Brexit referendum, to influence the election result and favor the politician they want to be elected (de Rochegonde & Tenenbaum, 2021). Hammond-Errey's (2019) study mentioned that Russia has a long history of the cyber influence and is the most advanced country in disinformation.

According to an exploratory study (Henschke et al., 2020), all those manipulations surrounding the political context affected negatively trust in the democratic institutions. Those authors mentioned that trust in government has significantly declined among democratic states. And inference in several elections is the cause of this declining situation and opens the door for more discriminatory discourse on social media to erode democratic states. Indeed, the more discrepancies in political discourse, the more openings for disinformation messaging and that will amplify distrust (Downes, 2018).

7.3.3 Economical Objective

Cyber threat actors use cyber influence to gain economic objectives (Government of Canada, 2018). Here are different examples.

According to a study from the Swedish Defense Research Agency (Larsson, 2006), Russia applied economic pressure concerning its natural gas exportation to eastern European countries by either cutting off supplies or selling prices. The study concludes that the more favorable a country's policies are toward Russia, the lower the prices and better terms of the contract are offered. Indeed, recently, Russia applied economic pressure concerning its natural gas exportation to European countries. Since April 27, Russia had cut the fuel supply to Poland and Bulgaria. Putin also asked to be paid in rubles. Noted that Russia is the largest European supplier of natural gas, so he has the power of influence.

As for China, this country uses cyber influence to serve its political and economic objectives as well. This country is well-known for cyber espionage. China's cyber activities target political, economic, military, educational organizations and so on to obtain illegally or legally sensitive US software and technology. In 2021, the US Intelligence Community assessed that the People's Republic of China presents a growing influence threat. The cyber-espionage operations have targeted telecommunications enterprises, providers of broadly used software, etc. (Government of US, 2021).

Another economic influence is the case of gun culture in the US Firearm manufacturers have played a major role in influencing American gun culture. Indeed, the more firearm manufacturers sell a gun, the more they make money (Siegel, 2018). A law against carrying guns is not in their favor. Therefore, the National Rifle Association (NRA) which represents those manufacturers is a powerful political machine. This association finances those who vote the laws, and those who are elected: in short, a corrupt box. According to Alhazbi (2020), the NRA organization had spent more than $30 million supporting the Trump campaign. It's not the first time, it's the way of doing things.

7.3.4 Ideological Objective

Hostile foreign states want to promote their ideological interests, like climate change, abortion, COVID-19, etc. Here are some examples where cyber influence has been used for an ideological purpose.

In 2019, the Alberta election was at risk of cyber influence because of environmental issues. The pipeline was a divisive issue and a perfect target for manipulating elections (Gouvernement du Canada, 2019). Indeed, accounts affiliated with lobbying groups posted false messages on social media. Those accounts have promoted separatism ideology (official website: https://albertaindependance.ca).

New data published by researchers at the University of Sherbrooke show that more and more people are hesitant about the COVID-19 vaccine. The World Health Organization has mentioned that one of the biggest threats to public health is vaccination hesitation. The movement anti-vax is very active on social media, and it generates billion of revenue. In fact, in January 2021, one of the largest vaccination centers, Dodger Stadium in Los Angeles, was shut down after antivaccine protestation over Facebook.

In 2018, the Irish voted to ban illegal abortion. And it's illegal for foreign entities to contribute to political elections. However, 14 US campaigns influenced abortion choice. Despite Facebook and Google's political advertisement restrictions from outside of Ireland, foreign advertisements were still published and promoted anti-abortion (Haley Ott, 2018).

For the years to follow, more cyber influence on climate change will circulate on the social network. Conspiration movements are very active. As we already mentioned, there are influencers and megadonors like Robert Mercer, who ejected the scientific consensus on climate change.

7.4 CYBER INFLUENCE ON SOCIAL MEDIA

In this section, we will understand why social media are so popular as a cyber tool to influence a mass of people, and we will explain a little bit how it works.

7.4.1 Reasons for Its Popularity

Several studies have shown that social networks are very effective to spread disinformation rapidly to a lot of people and therefore make it easier to influence individuals' opinions and behaviors (Badawy et al., 2018). Another reason for its social media popularity is that it's cheap and simple. Russia's IRA was able to reach an estimated 120 million Americans during the US 2016 presidential election, with advertisements that cost only $100,000 (Henschke et al., 2020).

This trend is increasing over time. Indeed, the number of states that use cyber influence on social media has increased from 28 states in 2017 to 70 in 2019 (Boily, 2021). It's almost quadruple in two years!

Furthermore, viral fake news can generate a lot of dollars in advertising revenue. However, false publications of fact concerning a public figure (e.g., celebrity or government official) are actionable for defamatory (Klein & Wueller, 2018).

According to Carnegie Endowment for International Peace research, among all the social media platforms, the most used by users are Facebook and Twitter. During the survey from 2015 to 2020, in 54% of cases, Facebook and Twitter are the user's preferences (Yadav, 2021). But some other platforms like TikTok have grown substantially. It would be interesting to investigate how threat actors use TikTok to make cyber influence? Is it easier to use other conventional platforms, like TikTok, Telegram, and Reddit to cyber influence?

7.4.2 How is Disinformation Spreading so Fast and Wildly on Social media?

Disinformation is hard to perceive for humans. In general, people tend to like, share, and react to this content because they think it is true. They share with their friends and all their cluster that it's easier to influence. Furthermore, research indicates that repeating the same fake news ideas increased the probability that humans will believe it (Grimes, 1978).

Cyber threat actors can create a false impression that millions of people share the message using automated tools and techniques, like bots and trolls. The disinformation will spread if the users have a lot of followers, such as celebrities or politicians. This author mentioned that "one of the most prolific fake accounts during the 2016 election was the official Twitter account of the Tennessee Republican Party @TEN_GOP with over 130 million followers" (Cunningham, 2020).

Like humans and bots, algorithms help to spread disinformation. Indeed, algorithms recognize the users' engagement (like, retweet, reaction users). For example, as reported in Sinan's book, Elon Musk has more than 40 million Twitter followers and by adding #bitcoin to his post and tweeting about dogecoin, he has increased cryptocurrency market events (Sinan, 2021).

That's scaring how algorithms can move and spread cyber influence rapidly. And without knowing what's happening. It's out of control! Fake news start, take different ways to spread and we don't know when or whether it will end or not. It's a huge spaghetti which takes several servers and thus consumed a lot of energy unnecessarily.

7.5 TECHNIQUES OF CYBER INFLUENCE

In this section, we will discuss the set of techniques used to influence people on social media. The most techniques used by hostile actors are, by order of importance:

1. Trolls and bots;
2. Fake accounts;
3. Hijacking existing social accounts/hashtags;
4. Deepfakes.

According to this study made during 2013–2018 (Martin & Shapiro, 2019), trolling techniques are the most used 90% of the time. 50% of the influence is made with an automated message. Trolling, bots, and hashtag hijacking were used together 97% of the time. To understand what we are talking about, here are definitions of each term.

7.5.1 Trolls and Bots

Trolls can be defined as a human that aims to generate controversy, unlike bots which are robots that emulate the activity of human users. A mix of robots and humans exists as well to make the online conversation more real.

The advantage of using automated tools is the fastest spreading disinformation on social media. Research indicates that Russian cyber influence used bots and trolls, to modify the opinions of citizens on popular sites such as Twitter, YouTube, and Facebook. Indeed, the Russian Army Troll Farm represents the first revealed case of hiring human operators to carry out a deceptive online interference campaign (Badawy et al., 2018).

Bots are estimated to comprise 9–15% of all Twitter accounts (Varol et al., 2017). And Twitter users are over 326 million. It results that 10% of bots totalize 32.6 million bots. That's huge and difficult to overcome and get rid of.

7.5.2 Fake Accounts

Facebook's threat Report 2021 on combating influence operations details nearly 150 disinformation operations. They found a lot of fake accounts, completely created characters, used for example, as journalists or as residents of the targeted countries, like hiring people from Ghana as if they were Black Americans discussing politics and racial issues or hiring Mexican people to have an online discussion on Hispanic pride and the Black Lives Matter movement. One of their strategies was to attract people to their sites to promote fake contents (Boily, 2021).

In 2018, the US Department of Justice accused 13 Russians of interfering in the US 2016 presidential election, plus three Russian entities, including IRA. The purpose of the charges is based on stolen identities and fake social media accounts. This fake operation was backed by a Kremlin associate, Prigozhin, known as "the chef" who is also accused (Henschke et al., 2020).

7.5.3 Hijacking Existing Social accounts/hashtags

Hijacking is stealing and taking an existing identity. Threat actors then take your social media account and send a message on your behalf. For example, Americans' identities have been stolen to create PayPal accounts. Indeed, the black-market prices for hacked PayPal accounts have nearly tripled during the pandemic. Recently, PayPal admitted that 4.5 million accounts were fake.

In recent years, digital activism appeared like the hashtag #BlackLivesMatter, which became a unifying theme of racial injustice (Yang, 2016). Furthermore, hashtags can be used to spread fake news. Let's see this example cited in Cunningham's book. In 2017, Syrian fighter jets dropped chemical munitions on the Syrian town and injured and killed over 300 people. #SyrianGasAttack hashtag was the true story and the #SyriaHoax hashtag was the fake news. It has resulted that the #SyriaHoax hashtag spreading faster and broader than the truth and, in consequence, had reached more people (Cunningham, 2020).

Moreover, hashtags sent by trolls were highly important during US 2016 election period but decreased thereafter. Hashtags also spread fake news exponentially. According to this study, hashtags sent by trolls were highly important during US 2016 election period but decreased thereafter. They supposed that IRA realized that hashtags could identify their activities (Alizadeh et al., 2020).

7.5.4 Deepfakes

Deepfakes are a way of transmitting messages without any text. That's an advantage. Furthermore, it's faster than other communication methods because there is no need to read (Zannettou et al., 2019). With face-swapping applications, such as FaceApp, SnapChat, MixBooth, FaceBlender, and many more, it's easy to replace a face with another one (Kumar et al., 2020).

Major social media have published policies forbidding the use of deepfakes for disinformation purposes. And they invested resources in that domain. For example, in 2021, Facebook started to work with Michigan State University researchers to develop a system that can detect and trace deepfake source (Sarwar, 2021).

Governments have also acted to accelerate research in the domain, such as the Defense Advanced Research Projects Agency. The Identifying Outputs of Generative Adversarial Networks Act would direct government research resources toward advancing the detection of manipulated media (Hwang, 2020). Indeed, experts are worried our adversaries would use the deepfake technology to meddle in an election (Mark & Temple-Raston, 2020). According to Dobber et al. (2021), deepfakes can be impactful along with political microtargeting techniques, which consist of targeting personalized advertisements to each person.

7.6 COUNTERMEASURES OF CYBER INFLUENCE

In this section, we will discuss the most used countermeasures by social media since 2019. We will also discuss some techniques used to detect disinformation. Therefore, we made a resume on some related work on trolls' and bots' detection as well as deepfakes.

7.6.1 Countermeasures Used by Social media

Countermeasures can be preventive or detective. Preventive measures are manual and automated controls that can eradicate fake news like accounts' and videos' authentication. Detective measures are manual and automated controls that permit to detect, after fake news have spread on social media like labeling and fact-checking. So, preventive measures are better. However, both, preventive and detective, are necessary to be implanted to have a better countermeasure result.

Social media have put in place measures to counter cyber influence impacts during the last two years. Several factors may be driving these trends, especially since COVID-19 and the US 2020 election.

However, social media platforms are criticized for focusing on specific threats (Yadav, 2021). Indeed, since 2016, Facebook publishes regular reports related to specific disinformation topics, such as elections, climate change, Russia's invasion, COVID-19, and so on (Meta, 2021b). As of Twitter, they publish their own regular reports on their specific's disinformation topics as well (Twitter, 2019).

That's a real matter here. Social media platforms try to extinguish fires. There is no proactive method. They're banning fake accounts that already spread fake news. The conspiracy movement and other influencers take advantage of this situation.

The message promoted by social media platforms seems that they don't like to take the responsibility and the risk against liberty speech which contravenes the democratic culture. Social media users have also the responsibility to limit the spreading of fake news. They must consult and make a judgment on what they read before liking or writing fake news comments.

We present here a survey made by the Carnegie organization related to countermeasures used by social media for 104 interventions during 2019–2020. Redirection and labeling/fact-checking are the most used 77% of the time. Those solutions are preferred over banning troll and bot accounts (Yadav, 2021). We will explain a little bit the definition of redirection, labeling, and fact-checking.

7.6.1.1 Redirection

Redirection consists of inviting users to click on the link to redirect content on the official website. According to Carnegie, redirection was popular due to the COVID-19 pandemic: several platforms have redirected users to information from the World Health Organization and the US Centers for Disease Control and Prevention.

7.6.1.2 Labeling

Content labeling consists of adding context to prevent users from possible fake posts or identifiers like advertisement funding disclosures or article publication dates. For example, labels "! get the fact about COVID-19" are shown on posts containing information related to COVID-19.

This study mentioned that labels cannot be efficient to advise users if they are placed in a way that users will miss. For them, the better place to put the label is right in the middle of the video or message (Nassetta & Gross, 2020).

7.6.1.3 Fact-checking

Fact-checking is the verification of the content. Social media collaborate with a third-party organization to stay impartial about what's true or false information and they have limited resources. In fact, according to Facebook's website, they collaborate with independent third-party verification media certified by the nonpartisan International Information Verification Network (IFCN) (Meta, 2021a).

Since 2019, Twitter fact-checks people related to politics to instill confidence that the content is reliable and accurately reflects candidates' and elected officials' positions and opinions (Twitter, n.d.). But fact-checks alone are not enough to correct misinformation if fact-checkers themselves are not agreed. Rich et al., (2020) mentioned that fact-checking dispute is counterproductive, and politicians' fact-checks may never work.

As bots send automated messages rapidly to a mass of people, fact-checkers work will not be able to get rid of fake news as their work is mostly manually. And top of it, fact-checks intervene after the users have identified the contents as fake news. However, this study noted that fact-checking can reduce the impact of disinformation as well as their willingness to share with others (Courchesne et al., 2021).

7.6.2 Detection of Trolls and Bots

There is a need to develop automated approaches to identify troll and bot accounts as well as fake news and deepfakes to mitigate their impacts on the political process and protect people from opinion manipulation. Studies in that domain can be useful for social media companies to upgrade their tools and for governments to make an investigation.

In table 7.1 and 7.2, we have summarized a review of related work to detect trolls and bots for the years 2019–2021 with Twitter's databases. Twitter released all the accounts and related content in 2016 to enable independent academic research and investigation. They include more than ten million tweets and more than two million images, videos, etc. (Gadde & Roth, 2018).

TABLE 7.1
Proposed Approaches on Detecting Trolls and Bots 2019–2021

no	1	2	3	4	5	6
Title of the study	Automatic detection of influential actors in disinformation networks, 2021 (Smith et al., 2021)	Behavior-Based Machine Learning Approaches to Identify State-Sponsored Trolls on Twitter, 2020 (Alhazbi, 2020)	TexTrolls: Identifying Russian trolls on Twitter from a textual perspective, 2019 (Ghanem et al., 2019)	Content-based features predict social media influence operations, 2020 (Alizadeh et al., 2020)	An Empirical Study of Machine learning Algorithms for Social Media Bot Detection, 2021 (Heidari et al., 2021)	Deep contextualized Word Embedding for Text-based Online User Profiling to detect Social Bots on Twitter, 2020 (Heidari et al., 2020)
Objective	Identification of trolls, bots, and users accounts with high impacts	Identification of Saudi trolls' accounts regardless of the content	Identification of Russian trolls' accounts with textual features	Identification of (Russian, Venezuela, Chinese) trolls' accounts with features	Identification of bots' account, based on five categories of bots	Identification of bots' account, based on user's personal information
Data set	29 million tweets, 1 M accounts	Saudi Trolls: 1,681 accounts	2,023 trolls accounts with over 1,8 M tweets	2,660 Chinese trolls who post 1,9 M tweets, 3,700 Russian trolls who post 3,7 M tweets	Over 5,500 bots and 3,300 fakes followers who post over 6 M tweets	Over 5,500 bots and 3,300 fakes followers who post over 6 M tweets

(Continued)

TABLE 7.1 ***(Continued)***
Proposed Approaches on Detecting Trolls and Bots 2019–2021

no	1	2	3	4	5	6
Period of data collection	French election May 2007 to February 2020	December 2019	US election August to December 2016	January 2015 and December 2018	Cresci 2017 data set, a labeled data set of bots and human users on Twitter	Cresci 2017 data set, a labeled data set of bots and human users on Twitter
Features	Behaviors and content	Behaviors (number of tweets, retweets, hashtags, and URLs)	Emotional and linguistic	Post-URL pair (e.g., tweet containing a URL)	Sentiment features	User's personal information, such as age, gender, education, and personality
Result	96% precision Russian trolls	94.4% Saudi Trolls and 72.5% Russian	96% precision with all features	Average monthly score of 89% for the 47 months	87% for English and Dutch languages	Precision of 94% after multiple classification
Advantages	Impacts with specific narratives	Based on features from only 500 tweets, detect real time	Emotional features are independent of themes (election US)	Any platform. Running each task over multiple short time periods. Test over the longest feasible period (36 months for most tests)	Extraction of new sentiment features for bot's detection	A new bot detection model that outperforms previous bot detection techniques

Each method uses machine learning, like a classification algorithm to extract text content (e.g., sentiment analysis) or social content (e.g., number of retweets). Then, studies use other machine learning techniques (e.g., Random Forest) to suggest if it is a bot or human account.

All those studies have their pros and cons. Let's see in more detail two studies cited in the table that brings interesting added value.

A previous study (Smith et al., 2021) identified trolls, bots, and users with a high impact of influence, based on spreading specific narratives (e.g., hate Clinton). It showed a mapping of the influence narrative network where accounts of high impact are identified. Also, this model identifies both trolls and real users, like user @JackPosobiec and troll @TEN GOP, involved in the spread of disinformation.

Another previous study (Alhazbi, 2020) was based on features from only 500 posts, to classify the account if it is a troll or not. Because few posts are needed, disinformation will be detected in real time. Moreover, this general model is capable to identify these trolls without the need to analyze the contents of their post, which requires using natural language processing that will vary from one language to another.

All those studies proposed detecting tools to eliminate disinformation from trolls and bots. It's less effective than preventive measures though.

7.6.3 Deepfakes' Authentication and Detection

We summarized below a short review of related work on deepfakes. One study is based on an authentication method and the two others are detection methods.

The goal of authenticated video or image is to track the provenance and the history of digital content to its source. It proves that the video or image is authentic, and not fake. It's what we call "blockchain technology." Cryptocurrency is based on this

TABLE 7.2
Proposed Approaches on Authenticating and Detecting Deepfakes 2018–2019

no	1	2	3
Title of the study	Combating deepfake videos using blockchain and smart contracts, 2019 (Hasan & Salah, 2019)	Exposing Deepfakes Using Inconsistent Head Poses, 2019 (Yang et al., 2019)	Deepfake video detection using recurrent neural networks, 2018 (Güera & Delp, 2018)
Objective	Decentralized Proof of Authenticity system using the technology blockchain using Ethereum smart contracts	Solution for the detection of fake video	Solution for the detection of fake video
Dataset	No tests	241 real images and 252 deepfakes	600 videos that a half are fakes
Period of data collection	NA	2019	2018
Features	NA	Support Vector Machines (SVM)-based method to detect deepfake videos: comparing the face landmarks between the real images and fake images	Two-stage analysis composed of a CNN to extract features and Recurrent Neural Network (RNN) to capture inconsistencies
Result	NA	84% of detection	97% of detection
Advantages	Authentication of source. No deepfakes	Features extracted from 68 landmarks of the facial region of 3D head	With less than 2 seconds of video, the solution can accurately predict fake video

technology like Ethereum digital money. It captures the history of digital transactions to preserve their integrity. Any manipulation attempt can be easily discovered by comparing it with the original content. As of today, there are no established methods to authenticate deepfakes (Hasan & Salah, 2019).

Regarding the detection of deepfakes, different methods have been developed as face detection landmarks (SVM) and convolutional neural network (CNN) with recurrent neural network (RNN). Each method uses machine learning, a technique from the field of artificial intelligence, to detect any kind of manipulation in images and videos. In general, deepfakes that are poorly made are easily spotted.

7.7 DISCUSSION

7.7.1 Improving Countermeasures

In 2021, the major social media platforms meet with US Congress related to the disinformation major problems. Politicians and social media companies did not come up with an issue to resolve problems (Courchesne et al., 2021).

We have seen that there is a place for improvement. Indeed, fact-checking doesn't resolve the problems, either by redirection, labeling, or banning fake accounts. This is because social media and users generate too much content, which then spreads at extraordinary speeds. Even if social media use fact-checkers, their work is mostly manual, and they rely also on a human to detect fake news. In most cases, fact-checkers review content only after humans noticed it.

Furthermore, as we noted, social media focus on specific threats and, therefore, some major events passed by. For example, for the Capitol's event on January 6, 2021, events were organized and amplified in peace on Facebook. Alerts from some employees were ignored. On January 7, Facebook decided to ban Trump's account from the platform. The reason Facebook gave is that the team working on the election has been moved for other activities. Here is a good and recent example that fact-checkers or other measures don't work to ban fake news and there are not enough automated algorithms in place to go through cyber influence.

7.7.2 Cyber Influence Impacts

We will detail some major impacts caused by cyber influence. In sum, there are political, health, and economic impacts.

According to an Axios survey, only 55% of Americans believe Biden won the 2020 US election. As we already mentioned, trust in democracy is declining. On January 6, 2021, five people died. Trump and several Republican members repeated unfounded claims of electoral fraud affecting the 2020 election outcome. This study mentioned that when social problems exist, conspiracies take place (Andersen, 2021). We are still not sure what the truth is. Indeed, in June 2021, cyberattacks and fraud on electronic vote machines are on the way to being confirmed and several US states have recounted the votes.

According to this interview (Marist Poll, 2020), disinformation or misleading information, followed by voter fraud, are the most threats to election integrity, as shown by the graphic below. Indeed, we can conclude that fake news is, by far, the biggest threat to election integrity. But also, as we will see that fake news impact negatively other domains which are quite disruptive.

Related to the COVID-19 pandemic, the Council of Canadian Academies has noted that disinformation about masks, treatments, and vaccines has undermined public trust, sowed confusion, and affected the virus containment. In some places around the world, vaccine hesitancy has threatened herd immunity, causing infectious diseases and re-emerging measles (Council of Canadian Academies, n.d.). In fact, in January 2021, one of the largest vaccination centers, Dodger Stadium in Los Angeles, has been shut down after antivaccine protestation over Facebook. Adherence to conspiracy theories and the existence of the antivaccine movement would have influenced this trend (Noël, 2021).

In regard to stock price manipulation, the investigation which has been made by US Securities and Exchange Commission validated the link between some fake news and stock price movements. In 2014, SEC filed a lawsuit against several companies and related fake news companies, such as Lidingo and DreamTeam.

7.8 CONCLUSIONS AND FUTURE WORK

In this work, we have seen that cyber influence is defined as disinformation or fake news spreading on social media platforms. Threat actors, mainly major countries, use influence to accomplish their strategic goals: economic, political, or ideologic.

Techniques used by threat actors are concentrated on using robots, and some studies demonstrated that those bots spread more and faster fake news than humans do. Gartner predicts "that most individuals in developed economies will consume more false than true information by 2022" (Gartner, 2019). Facebook and Twitter are the two most used platforms to do so but some others like TikTok expands substantially, where it gave place to fake news. Even out of the election period, fake news is still high. Their algorithms, based on users' engagement (likes, retweets, etc.), make that happen. Fake news is spread faster by influencers who are known to be big influencer players. That will give the chance to expand bad news.

We can conclude that disinformation erodes trust in information: people are not able to differentiate true stories from false ones. Social media have been proven as effective tools to cyber influence individuals' opinions and behaviors. Impacts are huge: from harm to people's health to violence, from economic instability to increase social problems.

In the future, a possible improvement would be government regulation to compel more algorithm transparency. The algorithm is a black box, we don't know how it works, and why it came with this result. There is no explanation. In that way, users will be more aware of their choices, and make clearer and more rational decisions.

The second way of improvement would be tracking who are the influencers and those who are acting undercover like megadonors. The bad influencers can like, comment, and then spread fake news all over social media. It would be interesting to analyze which social media platform is more willing to spread fake news as well.

REFERENCES

Alhazbi, S. (2020). Behavior-based machine learning approaches to identify state-sponsored trolls on Twitter. *IEEE Access*, *8*, 195132–195141.

Alizadeh, M., Shapiro, J. N., Buntain, C., & Tucker, J. A. (2020). Content-based features predict social media influence operations. *Science Advances*, *6*(30), eabb5824.

Andersen, T. (2021). Before mob stormed US Capitol, Trump told them to 'fight like hell'. *The Boston Globe*. https://www.bostonglobe.com/2021/01/06/metro/heres-what-trump-told-his-supporters-before-many-them-stormed-capitol-wednesday/

Badawy, A., Ferrara, E., & Lerman, K. (2018). Analyzing the digital traces of political manipulation: The 2016 Russian interference Twitter campaign. In *2018 IEEE/ACM International Conference on Advances in Social Networks Analysis and Mining (ASONAM)*, (pp. 258–265). IEEE, Barcelona, Spain.

Berge, G. (2018). Foreword. In C. Ireton & J. Posetti (Eds.), *Journalism, fake news & disinformation*. Paris: UNESCO. https://unesdoc.unesco.org/ark:/48223/pf0000265552

Boily, A. (2021, mai 28). Selon Facebook, la Russie est le principal producteur de désinformation. *Journal de Montréal*. https://www.journaldemontreal.com/2021/05/28/selon-facebook-la-russie-est-le-principal-producteur-de-desinformation

Cadwalldr, C. (2018, April 7). Facebook suspends data firm hired by Vote Leave over alleged Cambridge Analytica ties. *The Guardian*. https://www.theguardian.com/us-news/2018/apr/06/facebook-suspends-aggregate-iq-cambridge-analytica-vote-leave-brexit

Council of Canadian Academies. (n.d.). *The socio-economic impacts of science and health misinformation*. https://www.cca-reports.ca/reports/the-socio-economic-impacts-of-health-and-science-misinformation/

Courchesne, L., Ilhardt, J., & Shapiro, J. N. (2021). Review of social science research on the impact of countermeasures against influence operations. *Harvard Kennedy School Misinformation Review*. September 2021, Volume 2, Issue 5, 1–17 pp. https://misinforeview.hks.harvard.edu/wp-content/uploads/2021/09/courchesne_social_science_research_review_countrmeasures_20210913.pdf

Cunningham, C. (2020). *Cyber Warfare-truth, tactics, and strategies: strategic concepts and truths to help you and your organization survive on the battleground of cyber warfare*. Packt Publishing. Birmingham, UK.

Delevingne, L. (2014, November 8). Have Mercer! The money man who helped the GOP win. *CNBC News*. https://www.cnbc.com/2014/11/07/robert-mercer-the-most-important-political-money-man-youve-never-heard-of.html

Derosa, K. (2018, March 25). Victoria firm AggregateIQ denies link to data-miner at heart of Facebook controversy. *Times Colonist*. https://www.timescolonist.com/local-news/victoria-firm-aggregateiq-denies-link-to-data-miner-at-heart-of-facebook-controversy-4660626

de Rochegonde, L., & Tenenbaum, E. (2021). Cyber-influence: les nouveaux enjeux de la lutte informationnelle. https://hal-sciencespo.archives-ouvertes.fr/hal-03389162..

Dobber, T., Metoui, N., Trilling, D., Helberger, N., & de Vreese, C. (2021). Do (microtargeted) deepfakes have real effects on political attitudes? *The International Journal of Press/Politics*, *26*(1), 69–91.

Downes, C. (2018). Strategic blind–spots on cyber threats, vectors and campaigns. *The Cyber Defense Review*, *3*(1), 79–104.

Gadde, V., & Roth, Y. (2018, October 17). Enabling further research of information operations on Twitter. https://blog.twitter.com/en_us/topics/company/2018/enabling-further-research-of-information-operations-on-twitter

Gartner. (2019). Gartner top strategic predictions for 2018 and beyond. https://www.gartner.com/smarterwithgartner/gartner-top-strategic-predictions-for-2018-and-beyond

Ghanem, B., Buscaldi, D., & Rosso, P. (2019). TexTrolls: identifying Russian trolls on Twitter from a textual perspective. *arXiv preprint arXiv:1910.01340*. https://arxiv.org/abs/1910.01340

Gleicher, N., Franklin, M., Agranovich, D., Nimmo, B., Belogolova, O., & Torrey, M. (2021). Threat report: The state of influence operations 2017-2020. https://about.fb.com/wp-content/uploads/2021/05/IO-Threat-Report-May-20-2021.pdf

Government of Canada. (2018). *Cyber threats to Canada's democratic process*. Retrieved from https://cyber.gc.ca/en/guidance/overview-cyber-threats

Gouvernement du Canada. (2019). *Analyse des élections en Alberta*. https://www.international.gc.ca/gac-amc/assets/pdfs/publications/Analyse.elections.Alberta-FR.pdf

Government of Canada. (2020). *National cyber threat assessment 2020*. https://www.cyber.gc.ca/en/guidance/national-cyber-threat-assessment-2020

Government of Canada. (2021). *Cyber threats to Canada's democratic process: July 2021 update*. https://cyber.gc.ca/en/cyber-threats-canadas-democratic-process-july-2021-update

Government of US. (2021). *Chinese cyber threat overview and actions for leaders*. https://www.cisa.gov/sites/default/files/publications/CISA_Insights-Chinese_Cyber_Threat_Overview_for_Leaders-508C.pdf

Grimes, A. J. (1978). Authority, power, influence and social control: A theoretical synthesis. *Academy of Management Review*, *3*(4), 724–735.

Güera, D., & Delp, E. J. (2018). Deepfake video detection using recurrent neural networks. In 2018 15th IEEE international conference on advanced video and signal based surveillance (AVSS) Nov 27 (pp. 1–6). IEEE, Auckland, New Zealand.

Haley Ott, J. (2018). Social media under scrutiny ahead of Ireland abortion vote. *CBS News*. https://www.cbsnews.com/news/ireland-abortion-referendum-foreign-influence-groups-on-social-media/

Hammond-Errey, M. (2019). Understanding and assessing information influence and foreign interference. *Journal of Information Warfare*, *18*(1), 1–22.

Hasan, H. R., & Salah, K. (2019). Combating deepfake videos using blockchain and smart contracts. *IEEE Access*, *7*, 41596–41606.

Heidari, M., James Jr, H., & Uzuner, O. (2021). An empirical study of machine learning algorithms for social media bot detection. In *2021 IEEE International IOT, Electronics and Mechatronics Conference (IEMTRONICS)*, Toronto, ON, Canada.

Heidari, M., Jones, J. H., & Uzuner, O. (2020). Deep contextualized word embedding for text-based online user profiling to detect social bots on twitter. In 2020 International Conference on Data Mining Workshops (ICDMW), Sorrento, Italy.

Henschke, A., Sussex, M., & O'Connor, C. (2020). Countering foreign interference: election integrity lessons for liberal democracies. *Journal of Cyber Policy*, *5*(2), 180–198.

Hwang, T. (2020). *Deepfakes - Primer and Forecast - stratcomcoe*. https://stratcomcoe.org/publications/deepfakes-primer-and-forecast/42

Klein, D. O., & Wueller, J. R. (2018). Fake news: A legal perspective. *Australasian Policing*, *10*(2), 5–13.

Kumar, A., Bhavsar, A., & Verma, R. (2020). Detecting deepfakes with metric learning. In *2020 8th International Workshop on Biometrics and Forensics (IWBF)*, Apr 29 (pp. 1–6). IEEE, Porto, Portugal..

Lachapelle, J. (2022, 17 janvier). *Truth Social un pas en avant pour la désinformation*. https://www.tvanouvelles.ca/2022/01/17/truth-social--un-pas-en-avant-pour-la-desinformation

Larsson, R. L. (2006). *Russia's energy policy: security dimensions and Russia's reliability as an energy supplier*. Stockholm, Sweden: Swedish Defence Research Agency (FOI).

Laviola, E. (2018). QAnon conspiracy: 5 fast facts you need to know. *Heavy.com*. https://heavy.com/news/2018/08/qanon-conspiracy-trump/

MacKenzie, P. J. (2018). Cyberspace and cyber-enabled information warfare. In *Joint Air & Space Power Conference 2018*, Germany. https://www.japcc.org/wp-content/uploads/JAPCC_Read_Ahead_2018.pdf

Maria. (2022, 22 mars). Guerre en Ukraine: un champ de mines pour les hacktiviste. *No Stress News*. https://nostress.news/guerre-en-ukraine-un-champ-de-mines-pour-les-hacktiviste/

Marist Poll. (2020). *How the survey was conducted*. https://maristpoll.marist.edu/wp-content/uploads/2020/01/NPR_PBS-NewsHour_Marist-Poll_USA-NOS-and-Tables_Election-Security_2001140949.pdf

Mark, T., & Temple-Raston, D. (2020). Where are the deepfakes in this presidential election? *GPB News*. https://www.gpb.org/news/2020/10/01/where-are-the-deepfakes-in-presidential-election

Martin, D. A., & Shapiro, J. N. (2019). *Trends in online foreign influence efforts*. Working Paper. Princeton, NJ: Princeton University.

Meta. (2021a). *Fonctionnement du programme de vérification tierce de Facebook*. https://www.facebook.com/journalismproject/programs/third-party-fact-checking/how-it-works

Meta. (2021b). *October 2021 coordinated inauthentic behavior report*. https://about.fb.com/news/2021/11/october-2021-coordinated-inauthentic-behavior-report/

Nassetta, J., & Gross, K. (2020). State media warning labels can counteract the effects of foreign misinformation. *Harvard Kennedy School Misinformation Review*. https://misinforeview.hks.harvard.edu/article/state-media-warning-labels-can-counteract-the-effects-of-foreign-misinformation/#

Noël, B. (2021). Antivaccins: le doute à forte dose. *Radio-Canada*. https://ici.radio-canada.ca/recit-numerique/1838/antivaccins-trump-complot-pandemie-quebec

Rich, T. S., Milden, I., & Wagner, M. T. (2020). Research note: Does the public support fact-checking social media? It depends who and how you ask. *The Harvard Kennedy School Misinformation Review*. https://misinforeview.hks.harvard.edu/article/research-note-does-the-public-support-fact-checking-social-media-it-depends-who-and-how-you-ask/

Sarwar, N. (2021, June 16). How Facebook's new AI fights deepfakes by identifying creators. *Screenrant*. https://screenrant.com/facebook-deepfakes-source-creator-ai-detection-technology-explained/

Siegel, M. (2018). How the firearms industry influences US gun culture, in 6 charts. *Boston University*. https://www.bu.edu/articles/2018/firearms-industry-influences-us-gun-culture-in-6-charts/

Sinan, A. (2021). *The Hype Machine: How social media disrupts our elections, our economy, and our health and how we must adapt*. Milano, Italy: Currency.

Smith, S. T., Kao, E. K., Mackin, E. D., Shah, D. C., Simek, O., & Rubin, D. B. (2021). Automatic detection of influential actors in disinformation networks. *Proceedings of the National Academy of Sciences*, *118*(4), e2011216118.

Smith, V., & Thompson, N. (2020). *Survey on countering influence operations highlights steep challenges, great opportunities*. https://carnegieendowment.org/2020/12/07/survey-on-countering-influence-operations-highlights-steep-challenges-great-opportunities-pub-83370

Twitter. (n.d.). *Government and state-affiliated media account labels*. https://help.twitter.com/en/rules-and-policies/state-affiliated

Twitter. (2019). *Retrospective Review Twitter, Inc. and the 2018 Midterm Elections in the United States*. https://blog.twitter.com/content/dam/blog-twitter/official/en_us/company/2019/2018-retrospective-review.pdf

Varol, O., Ferrara, E., Davis, C., Menczer, F., & Flammini, A. (2017). Online human-bot interactions: Detection, estimation, and characterization. *Proceedings of the International AAAI Conference on Web and Social Media*, Montreal, Quebec, Canada.

Yadav, K. (2021). *Platform interventions: How social media counters influence operations.* https://carnegieendowment.org/2021/01/25/platform-interventions-how-social-media-counters-influence-operations-pub-83698

Yang, G. (2016). Narrative agency in hashtag activism: The case of# BlackLivesMatter. *Media and communication*, *4*(4), 13.

Yang, X., Li, Y., & Lyu, S. (2019). Exposing deepfakes using inconsistent head poses. In *2019 IEEE International Conference on Acoustics, Speech and Signal Processing (ICASSP)* Brighton, UK, 2019, pp. 8261–8265, doi: 10.1109/ICASSP.2019.8683164.

Zannettou, S., Bradlyn, B., De Cristofaro, E., Stringhini, G., & Blackburn, J. (2019). Characterizing the use of images by state-sponsored troll accounts on Twitter. Proceedings of the International AAAI Conference on Web and Social Media, 14(1), 774–785. 10.48550/arXiv.1901.05997

Postface

Schallum Pierre
Institute Intelligence and Data (IID),
Université Laval, Québec, Canada

Fehmi Jaafar
Department of Mathematics and Computer Science,
Québec University at Chicoutimi, Québec, Canada

CONTENTS

1. Quantum Computing and Ethics 176
2. NFT and Privacy 176
3. Industry 4.0/5.0 and Risk to Sensitive Data 177

Data is today at the center of the technological transformation of our societies. As an intangible asset, it differs from other goods such as oil with which it is often wrongly compared. Indeed, data does not dry up when we use it. Data's ability to be reused – its secondary use – ad infinitum puts it in the category of a "nonrival"[1] asset. These characteristics are essential to develop emerging technologies such as AI and blockchain as well as to create added value in societies. The gross domestic product (GDP) is the total monetary or market value of all the finished goods and services produced within in a specific year. According to PwC,[2] by 2030, a 14% growth in global GDP is expected from AI, or $15.7 trillion. Depending on their degree of preparation and involvement, some countries with low GDP today could benefit and experience significant advances. If AI could be an advantage for today's societies, the privacy issues related to its management must be raised in order to design eco-responsible innovations that are respectful of the human being and of the environment.

In the different chapters of this book, we explored the main domains that represent considerable risks for the respect of privacy, such as education, health, finance, or social media.

Through their place in the massive data production industry, the Internet of Things (IoTs) participates in the development of AI and is increasingly attracting the attention of web giants, governments and especially all types of hackers. Thanks to this book, private and public organizations will have at their disposal a tool that highlights on the one hand, the major challenges raised by privacy in the context of the Internet of Things and on the other hand recommendations for improving good practices.

Digital identity is presented as a bulwark for the protection of privacy. It opens up new avenues for improving digital trust. Concretely, there are a set of challenges

DOI: 10.1201/9781003227656-12

that are associated with the management of digital identity mainly in relation with the compliance and governance of personnel data in order to eliminate privacy and security risks.

However, if the issue of privacy, raised by ethics and dealt with in this book, is mostly associated with an individual, it is also imperative to deepen the risks on a societal and environmental scale. Indeed, the societal and environmental aspects still remain open from a research perspective. It would be relevant to explore the impact on society of the development of:

- Quantum computing;
- NFTs; and
- Industry 4.0.

1. QUANTUM COMPUTING AND ETHICS

Although quantum computers are still in the stage of emerging technology, it is important to anticipate its potential impact on privacy. Quantum computers[3] use qubits (quantum bits) instead of binary digits (bits) as in traditional computers. A qubit can exist at both 0 and 1, unlike a bit that exists only at 0 or 1. Thus, computers using qubits will have the ability to encode much more information and solve long, complex problems in a shorter time. At this rate, when quantum technologies become mature, they will be able for example to brute-force crack the encryption algorithms used today. The ethical question must be central in the design of quantum projects. From now on, it is important to anticipate the form that postquantum cryptography[4] will take. What can the protocols of Ben-Or, Goldwasser, and Wigderson [BGW][5] do to secure quantum technologies? In fact, the challenge that quantum development represents is a new avenue for research in ethics.

2. NFT AND PRIVACY

Nonfungible tokens (NFTs) have become a synecdoche of W 3.0 based on the intellectual property of content. These unique digital identifiers that cannot be copied, substituted, or subdivided, are recorded in a blockchain. Until the advent of NFTs, the inability to protect the intellectual property of a digital creation was a major obstacle to its integration into the art market and contributed greatly to its devaluation. Digital art, since its origins dating back to the 1960s, is thus difficult to monetize. In addition, it is difficult to ensure the uniqueness and authenticity of a digital work that can normally be shared and replicated ad infinitum. The authenticity of the work, the owner of the intellectual property, and the rarity or uniqueness of the digital work concerns an ecosystem composed of artists, collectors, institutional venues and people versed in technology/engineering, art history, finance and entrepreneurship. In other words, the digital artist must have at his disposal digital traces that can prove his authenticity to a collector or an institution that wishes to acquire his work, which can be put at the service of a person versed in art history. It is precise to all these questions that the crypto-art which uses the blockchain answers. On March 11, 2021, the work entitled "Everyday's: the First 5000 Days" by artist Mike Winkelmann, also known as Beeple, was acquired for USD 69,346,250 at Christie's.[6] The media coverage of this exceptional sale and, in the wake of it, that of the first Tweet by Jack Dorsey, the ex-chairman of Twitter,[7] has aroused the interest of many artists around the world, considering their vulnerable context during the COVID-19 pandemic.

Several privacy risks are associated with the use of NFT such as[8]:

- Anonymous social media harms while recovering online identifiers and avatars.
- The conformity of Blockchain with privacy laws.
- The management of location data.

These risks directly conflict with the spirit of many national and regional laws, by inference, such as the General Data Protection Regulation (GDPR) in Europe, the California Consumer Privacy Act (CCPA) in California and Bill 25 in Québec. NFTs could face significant penalties if they are found to be noncompliant. Significant technical, legal, and ethical challenges must be addressed before continuing to deploy these technologies through, for example, the meta-verse.

3. INDUSTRY 4.0/5.0 AND RISK TO SENSITIVE DATA

The value of data is crucial for the manufacturing industry, which must collect, store and process it. Sensors and IoTs facilitate the collection of massive data that allows the creation of predictive models and the operationalization of automation. Data-backed manufacturing in Industry 4.0 and 5.0 can improve all or part of its supply chain and bottom line. The need to collect more and more data increases the risk of cyberattacks.[9] Several governments and even hacker groups are working to gain access to the strategic data of large companies, universities, and governments working in areas such as biotechnology, aeronautics, university research. Work is to be considered from the point of view of cyberattacks dedicated to IoT, issues related to the presence of sensors in public places[10] and consent.

Other fields that use massive data deserve the attention of researchers from the point of view of privacy, such as autonomous cars or electric cars. Similarly, open databases that are made available to industry and research will need to be audited regularly to avoid potential misuse through re-identification. Finally, further research on DNA[11] data storage could prove to be of great benefit to the environment.

NOTES

1. Edmond Baranes, "La donnée numérique: un bien économique comme les autres?", *Cahiers français*. N° 419. *Le règne des données*, p. 20. https://www.vie-publique.fr/sites/default/files/2021-03/9782111574199.pdf, accessed on September 29, 2022.
2. PwC, *The macroeconomic impact of artificial intelligence,* Février 2018, p. 3. https://www.pwc.co.uk/economic-services/assets/macroeconomic-impact-of-ai-technical-report-feb-18.pdf, accessed on October 5, 2022.
3. Laurent Adatto, Fehmi Jaafar Schallum Pierre, "Cybersecurité, technologies quantiques, intelligence artificielle: le contexte pandémique alerte", *Alternatives économiques*, Septembre 16, 2021. https://blogs.alternatives-economiques.

fr/reseauinnovation/2021/09/16/cybersecurite-technologies-quantiques-intelligence-artificielle-le-contexte-pandemique-alerte, accessed on October 8, 2022.
4. Debbie Heywood, "Quantum computing – the biggest threat to data privacy or the future of cybersecurity?", *Taylor Wessing*, March 14, 2022. https://www.taylorwessing.com/en/interface/2022/quantum-computing---the-next-really-big-thing/quantum-computing---the-biggest-threat-to-data-privacy-or-the-future-of-cybersecurity, accessed on October 7, 2022.
5. Evan Koblentz, "New methods can protect data privacy against quantum computing", *New Jersey Institute of Technology*, October 13, 2021. https://news.njit.edu/new-tricks-can-protect-data-privacy-against-quantum-computing, accessed on October 7, 2022.
6. Christie's, *Beeple (b. 1981) everyday's the first 5000 days.* https://onlineonly.christies.com/s/beeple-first-5000-days/beeple-b-1981-1/112924, accessed on October 7, 2022.
7. Kevin Shalvey, "Twitter CEO's first tweet is expected to sell for $2.5 million. This is why the NFT is so valuable", *Business insider South Africa*, March 22, 2021. https://www.businessinsider.co.za/twitter-ceo-jack-dorsey-sell-first-tweet-nft-sunday-2021-3?r=US&IR=T#:~:text=Twitter%20CEO%20Jack%20Dorsey%20to,an%20NFT%20on%20Sunday%3A%20Explainer, accessed on October 7, 2022.
8. Michael Jacobs & Matthew Murphy, "NFTs: Privacy issues for consideration", *Locke Lord LLP*, January 27, 2022. https://www.jdsupra.com/legalnews/nfts-privacy-issues-for-consideration-7804114/, accessed on October 7, 2022.
9. Bryan Christiansen, "How safe is your data in Industry 4.0?", *EP&T*, August 6, 2019. https://www.ept.ca/features/how-safe-is-your-data-in-industry-4-0/, accessed on October 7, 2022.
10. République française, *L'internet des objets: vers de nouveaux rapports humains-machines.* https://www.vie-publique.fr/en-bref/283970-internet-des-objets-rapports-humains-machines/, accessed on October 7, 2022.
11. Sorbonne Université, *Le stockage des données sur l'ADN: une technologie révolutionnaire. Interview croisée avec Stéphane Lemaire et Pierre Crozet*, novembre 23, 2021. https://www.sorbonne-universite.fr/actualites/le-stockage-des-donnees-sur-ladn-une-technologie-revolutionnaire, accessed on October 7, 2022.

Index

A

Abbott, Tony, 3
Access control, 59, 78, 131, 134, 136, 139
Accountability, 41, 54, 89, 111, 113
Adaptation, 46, 49, 63, 64, 114, 116
AI act, 4, 109
Algorithms, 27, 47, 80, 82, 103, 105, 109, 162, 166, 169, 170, 176
Analytics, 46, 47, 53, 57
Anonymity, 14, 40, 52, 87, 105
Application layer, 9, 18, 20
Artificial intelligence (AI), 1, 39, 47, 48, 108
Asset(s), 51, 134
Australia, 3, 10, 17, 100; *see also* Australia Post Office
Australia Post Office, 17
Authentication, 2, 11, 12, 15, 17, 20, 30, 31, 59, 71, 74, 79, 80, 82, 83, 85, 87, 89, 103, 128, 131, 155, 164, 168
Authorization(s), 21, 26, 31, 78, 87, 105, 131
Automated decision-making, 114, 116
Automated Market Marker (AMM), 136
Autonomy, 27

B

Backdoors, 2
Bandwidth, 47, 80
Banks, 73, 80
Bannon, Steve, 156, 158
Behavioral, 41, 62; *see also* Behavioral data
Behavioral data, 75
Best practices, 13
Bill C-27, 4, 98, 101, 108, 109, 117
Biometrics, 2
Bitnation Refugee Emergency Response (BRER), 17
Blockchain, 1–4, 9–11, 13–21, 26–28, 30–34, 41, 56–58, 73, 85–87, 90, 127–129, 143–144, 168, 175–177; *see also* Blockchain Emergency ID (BE-ID)
Blockchain Emergency ID (BE-ID), 17
Blockchain layer, 20
Bluetooth low energy (BLE), 46, 74
Bot detection, 166–167
Botnet, 40, 55
Bots, 156, 162–170; *see also* Bot detection
Breaches, 12, 14, 21, 41, 60, 72, 84, 126, 142
Business, 2, 10, 13–16, 28, 33–34, 41, 44–46, 50, 54, 61, 63, 110, 127, 130; *see also* Business data
Business data, 2
Byzantine fault tolerance (PBFT), 13, 129

C

California Consumer Privacy Act (CCPA), 54, 177
Cambridge Analytica, 53, 156, 158
Canada, 64, 73, 88, 99–102, 104, 106–110, 113–114, 116–117, 157, 159–160; *see also* Canada's Consumer Privacy Protection Act
Canada's Consumer Privacy Protection Act, 64
Cardossier, research project, 28, 34
Certification, 28, 58, 128, 143
China, 10, 64, 98, 117, 157, 160; *see also* China's Personal Information Protection Law (2020)
China's Personal Information Protection Law (2020), 64
Citizen, 1, 72, 86, 88–89, 117–118, 163
Clearview AI tools, 98
Cloud computing, 46, 60, 78–80
Collaboration, 85
Communications Security Establishment of Canada, 157
Compliance, 11, 33, 63, 111–112, 129, 140, 176
Compute-to-data, 134, 136–137, 140, 154
Confidentiality, 2, 53–54, 85
Consensus, 13, 20, 57, 128–129, 161
ConsenSys, 17
Consent(s), 3, 14, 27, 41, 53, 61–64, 88, 101, 105, 107, 117, 126–127, 129, 138, 141–142, 144, 177
Consumer protection, 113
Convolutional neural network (CNN), 169
Creative commons, 62
Credentials, 11–12, 26, 28, 30–31, 33, 50, 77, 129–130, 139–140
Cryptocurrency, 13, 143, 162, 168
Cryptography, 13, 73, 82–83, 86–87, 129, 133, 176
Cyber command, 156
Cyber espionage, 160
Cyber influence strategies, 157
Cyberattacks, 40–41, 49, 58–60, 63, 158–159, 169, 177
Cybersecurity, 49, 58, 72–73, 100, 115–116

D

Data accuracy, 54
Data anonymization, 109, 111–112
Data breaches, 12, 21, 60, 72, 126
Data collection, 46, 54, 61, 72, 76, 88, 104, 116, 168
Data consumer, 129, 134, 136, 138–140, 143
Data custody, 139
Data double spending, 4, 125, 127–128, 137, 140, 142–144
Data injection, 49
Data leaks, 26
Data marketplace, 4, 125, 128, 133, 137, 140–144
Data minimization, 54
Data privacy, 4, 18, 21, 32, 42, 53, 57, 61, 63–64, 129, 143
Data processing, 4, 40, 54, 64, 107, 125, 128, 137, 142–144
Data protection, 4, 33, 41, 51, 62, 64, 72–73, 75, 83, 106
Data protection laws, 129
Data provider, 133–143
Data retention, 54
Data sharing, 12, 13, 127, 129–130, 135, 142, 148, 153
Data storage, 31, 58, 106, 139–140, 177
Data subject, 54, 127, 129–130
Data token, 133–136, 138–142, 151–152
Data transfer, 45
Data usage, 53, 58, 60, 129, 141
Data usage control, 129
Database, 2, 13, 20–21, 56, 85, 106, 117, 133, 166, 177
Decentralization, 33, 85, 143
Decentralized architecture, 127, 139
Decentralized identifiers (DIDs), 129–130
Deepfakes, 14, 62, 164, 166, 168–169
Deletion, 86
Democracy, 169
Detection, 2, 4, 59, 77, 103, 123, 127, 164, 166–169
Digital art, 176
Digital certificate, 12
Digital divide, 80, 87
Digital Identification and Authentication Council of Canada (DIACC), 17
Digital identity, 3–4, 9–21, 27, 29–34, 82, 85, 128, 175–176
Digital identity in Myanmar, 15
Digital signature, 31
Digital trust, 73, 77, 84–85, 87, 175
Disclosure, 27, 56, 165
Discrimination, 4, 98–99, 103, 111, 113, 115
Disinformation, 156–157, 160–166, 168–170
Distributed ledger, 20, 129
DNA, 177
Dpos (delegated proof of stake), 13

E

E-commerce, 10, 15–16, 25, 72
E-government(s), 10, 15
E-wallets, 72
Eavesdropping, 49, 59, 86
Eco-responsible innovation, 175
Economic objectives, 160
Education, 100–101, 104–106, 160, 167, 175
Education technology market, 101
eIDAS regulation, 29
Elections, 158–161, 164
Electronic Health Record (EHR), 50
Encryption, 30–31, 58, 60, 74, 77, 79, 82–85, 87, 90, 129, 133, 139–140, 143, 176; *see also* Fully Homomorphic Encryption (FHE)
Environment, 177
Equality, 103, 105
Errors, 103, 115–116
Estonia, 10, 85; *see also* Estonian e-residency program
Ethics, 72, 75, 89, 142, 175–176
Estonian e-residency program, 10
European Commission's, 4, 109, 115
European Court of Justice, 107
European Data Protection Supervisor (EDPS), 72
European Digital Identity framework, 29
EU's General Data Protection Regulation, 129

F

Facebook, 3, 53, 156, 158, 161, 163–165, 169–170
Facial recognition, 52, 98, 102, 105, 108, 117
Fact-checking, 164–165, 169
Fairness, 52–54, 127–128, 137, 143–144
Fake news, 165–166, 169–170
Finance(s), i, 4, 16, 157–158, 160, 175–176
Financial data, 1, 74, 76, 86
Financial sector, 16, 72
Findy, 29
Fintech, 72
Fog computing, 60
Foreign influence, 156
Freedoms, 2
French National Agency for the Security of Information Systems (In French: Agence nationale de la sécurité des systèmes d'information or ANSSI), 73
Fully Homomorphic encryption (FHE), 129
Fundamental rights, 2, 103, 115

G

General Data Protection Law (LGPD), 64
General Data Protection Regulation (GDPR), 2, 53, 73, 106

Geolocation data, 1, 62
Germany, 15, 62
Gigabytes (GB), 1
Governance, 2, 11, 17, 25–27, 32–34, 62–63, 86, 113, 116, 122, 143, 176
Google, 3, 43, 48, 79, 107, 131, 135, 137, 139, 161
Government of Alberta, 16
Greece, 16
Gulf Cooperation Council countries, 16

H

Hacktivists, 157–158
Harvard Law Review, 3
Health, 1, 10, 13, 15–20, 77, 114–116, 126–129, 131, 133, 135, 137–139, 141–144, 161, 165, 169–170, 175
Health data, 1, 126–129, 138, 142, 144
High-impact systems, 108, 110–111, 114
Hijacking, 77, 162–163
Host card emulation (HCE), 75
Human, 2, 11, 16, 28, 33, 40, 42, 48–52, 58, 60–63, 102–103, 108–110, 115–116, 143, 156–157, 162–163, 167, 169–171
Human control, 103, 115
Human rights, x, 52, 115–116, 157
Hybrid systems, 103
Hyperledger Indy/Aries protocol, 4, 128, 142

I

Identity correlation, 141
Identity credentials, 140
Identity fraud, 10–12
Identity management, 10–12, 15–18, 21, 27, 31, 128
Identity provider, 11–12, 25–26, 31
Identity theft, 11–12
Immutability, 33, 127
India, 10, 17
Industrial data, 1–3
Industry 4.0, 175–177
Infrastructure(s), 4, 16, 18, 28, 33–34, 40, 46, 60, 72, 79, 85, 129–130, 134, 136
Instagram, 3
Integrity, 2, 13–14, 16, 20–21, 33, 54, 85, 101–102, 105, 108, 111, 138, 169–170
Internet, 10, 15, 25, 31, 41, 43, 45–46, 51–54, 61, 64, 74, 76–80, 86–87, 98–99, 106, 117, 157–158
Internet of Things (IoT), 1, 2, 4, 26, 34, 40–51, 55–64, 79, 175, 177; *see also* IoT devices
Interoperability, 13, 18, 27, 29, 33–34, 78
Interplanetary File System (IPFS), 58
Intervention, 48, 89, 102, 159, 165
Intrusion detection, 59
IoT devices, 1–2, 26, 28, 34, 40–41, 46–48, 54–55, 57–60, 62–63
IoT ecosystem, 42, 48, 58
IRA, the Internet Research Agency, 158

J

Jamming, 59
Justice, 107, 163; *see also* European Court of Justice; US Department of Justice
JWM (JSON Web Messages), 31

K

Knowledge, 32, 53, 75, 108, 110, 115, 126, 141; *see also* Zero-knowledge proof(s) (ZKP)

L

Latency, 46–48, 60
Law(s), 3, 53–54, 60, 63, 73, 82, 86, 89, 104–111, 113–114, 117, 122–123, 129, 140, 157, 159–160, 177; *see also Harvard Law Review*
Learning models, 47, 59
Legal, 4, 33, 41, 53, 61–62, 64, 72, 77, 88–89, 97, 101, 103–109, 113, 115, 117, 141, 157, 159–161, 177
Legislation, 4, 15, 105, 114, 122, 129, 143; *see also* EU's General Data Protection Regulation
Licensing, 129–130, 132, 149
Literature review, 14, 89
Logging, 102

M

Machine learning, 47–48, 59, 83, 104, 108–110, 166–167, 169
Machine-readable, 62
Malware, 49, 55, 59, 76, 77–78, 82–84, 158
Management, 3, 10–12, 15–18, 20–21, 27, 31, 43, 48, 50, 80, 85, 88, 104, 111, 122–123, 127–129, 175–177
Manufacturing industry, 177
Media Access Control (MAC), 78
Metadata, 30, 129, 134–136, 139–141
Microsoft Corporation, 27
Minimization, 54, 108, 114, 116; *see also* Minimization principle
Minimization principle, 106
Mobile, 4, 27, 32, 50, 56, 73, 76, 77–85, 87, 89, 90, 131–132; *see also* Mobile payment; Mobile payment systems
Mobile payment, 72–73, 74, 76, 79, 81–82, 87, 89
Mobile payment systems, 4, 72–73, 75, 76, 80–81, 84
Monitoring, 1, 4, 42, 46, 50, 55, 101, 103, 133
Multi-stakeholder, 18

Mutual authentication, 12, 79, 85, 87
My Alberta Digital ID, 16

N

National Commission on Informatics and Liberty (In French: Commission Nationale de l'Informatique et des Libertés or CNIL), 73
Nation-states, 157
Natural gas, 160
Network, 4, 14–15, 18, 20, 29, 31, 41–43, 46–49, 55–60, 73, 76–80, 85–86, 109, 129, 158, 161, 164–169; *see also* Convolutional neural network (CNN); Recurrent neural network (RNN)
Nodes, 14, 20, 56–58, 60, 76, 81, 128–129, 134–135
Non-discrimination, 103
Nonfungible tokens (NFTs), 138, 176

O

Ocean protocol (OP), 133
Open banking, 15, 89
Open data, 89
OrgBook BC, 29

P

Payment, *see* Mobile
Payment Card Industry (PCI), 73
PBFT (practical byzantine fault tolerance), 13
Peer-to-peer (P2P), 56, 58
Performance, 43–46, 56, 79, 81, 88, 100, 104, 112, 115–116
Personal area network (PAN), 46
Personal data, 72–73, 77, 88–89, 104–107, 128, 133
Personally identifiable information (PII), 52
Personal Information Protection and Electronic Documents Act (PIPEDA), 73, 109
PIMN (Platform Identity Management Netherlands), 17
Planetoscope, 1
Point of sale (POS), 73
Political actors, 157–158
Portability, 27
Pos (proof of stake), 13
Postquantum cryptography, 176
Pow (proof of work), 13
Prescriptions, 50
Pricing, 134, 136, 142
Privacy, 1–5, 11–12, 14–16, 18, 20–21, 31–34, 40–43, 45, 47, 49, 51–53, 55–59, 61–64, 72–73, 79, 83–84, 86–87, 89, 100, 103–105, 108–110, 117, 127–129, 133, 140, 142–143, 175–177
Privacy associations, 89
Privacy impact assessment, 103, 107
Privacy policy, 61–62, 105–106
Private life, 1, 3
Private sector, 18, 89, 104, 105, 108, 114
Processing, 2, 4, 10, 40, 46–47, 49, 53–54, 60, 62–64, 82, 88, 105, 107, 110, 116, 127–128, 133, 136–138, 142–144, 168
Processor, 42, 48, 78, 105
Profiling, 4, 41, 111, 113, 166
Proportionality, 105
Provisioning, 138–139
Public key infrastructure (PKI), 85
Public sector, 89, 104–105, 107
Purpose limitation, 54

Q

Quantum computing, 175–176
QR code, 32, 85

R

Radiofrequency identification (RFID), 43
Real-time data, 40
Recourse, 115
Recurrent neural network (RNN), 168–169
Redirection, 74, 76, 165, 169
Regulation(s), 2, 4, 12–13, 29, 33, 41, 51, 53, 60–64, 73, 99, 101, 105, 110–117, 170
Relay attack, 75–76, 79, 81
Resilience, 116, 129
Responsibility, 109, 112, 156, 165
Restrictions, 161
Retail, 43, 50–51, 56, 77
Risk management, 48
Risk-based uses, 117
Russia, 157–161, 163–164, 166–167

S

Safety, 16, 81, 123
Schmidt, Eric, 3
Secondary usage, 127, 140–141
Secure element (SE), 75
Security, 143; *see also* Communications Security Establishment of Canada; Cyber Command; United States National Security Agency
Self-sovereign identity (SSI), 15, 26
Sensors, 1, 4, 41–43, 47–51, 60, 177
Service Alberta, 16
Service provider, 75–76, 80, 82–83, 104–106, 131, 133
Smart cities, 13, 40, 50
Smart contracts, 14, 134–136, 138, 143, 168
Smart devices, 42, 44, 49–50, 63

Smart environment, 40–41
Smart homes, 48–49
Smart retail, 5, 56
Smartphones, 40, 74, 78, 84–85
SMS, 73, 76, 80, 89
Social media, 1–4, 30, 52, 156–166, 169–170, 175, 177
Societal impact, 108
Spoofing attacks, 58
Stakeholder, 15, 18, 80, 89
Standards, 1, 18, 26, 29–33, 62, 73
Storage limitation, 54
Supply chain, 43, 177
Surveillance, 49, 98, 100, 102–103, 105–106, 117
Sustainability, 114
Swedish Defense Research Agency, 160
Switzerland, 28, 32
Symmetric Key, 78, 81
System(s), 2, 4, 10–18, 20–21, 27–33, 40–47, 49–51, 55–59, 63, 72–90, 99–117, 127–130, 134–135, 137–139, 140, 164, 168

T

Text mining, 48
Third party, 13, 15, 58, 87, 101, 104, 107, 133, 139–140, 142–143, 165; *see also* Third-Party Payment Provider (TPP)
Third-Party Payment Provider (TPP), 82
Threat(s), 3, 4, 41, 59, 63, 72, 76–77, 80, 123, 128–129, 157–158, 160–164, 169–170; *see also* Cyber espionage
Throughput, 45, 58, 129
TikTok, 62, 156, 161, 170
Tokenization, 78, 82
Tokens, 131, 133–136, 138–139, 141–143, 176
Tracking, 4, 43, 50–52, 63, 77, 102, 170
Transactions, 10, 13–14, 16, 20, 53, 56–57, 72–73, 75, 77, 79–80, 84, 86–88, 128–129, 134, 136, 140, 169
Transfer, 4, 45–46, 53, 58, 60, 77–78, 84, 89, 106–107, 133, 139–140
Transparency, 14, 27, 29, 41, 53–54, 57, 60–61, 86, 108–109, 112–116, 170
Trump, Donald, 53, 156, 158, 160, 169; *see also* Bannon, Steve
Trust, 11–13, 16–18, 21, 26–27, 32–34, 58, 61–63, 73, 75, 77, 82–85, 87, 127, 129–130, 139–140, 142–144, 159–160, 170, 175
Trusted execution environment (TEE), 83
Twitter, 55, 156, 158, 161–167, 170

U

Ukraine, 158
United Kingdom, 10, 17
United States, 10, 17, 51, 72, 100, 106–107, 156
United States National Security Agency, 156
Unstructured Supplementary Service Data (USSD), 73
URI (Uniform Resource Identifier), 30
US Department of Justice, 163
US Homeland Security, 17
Users, 11–16, 26–29, 33, 41, 46, 51, 53–54, 64, 73, 77, 79–80, 84, 89–90, 115, 117, 127, 141, 156, 158, 161–163, 165–170

V

Verifiable credentials (VCs), 30, 129
Viral, 156, 161
Voice recognition, 61
Vulnerabilities, 2, 11, 41, 56, 77, 123

W

Wallet(s), 26, 28, 31–33, 72, 139
Wi-Fi, 41, 43, 45, 80
Wireless application protocol (WAP), 74
Wireless Transport layer security (WTLS), 74
World Wide Web Consortium (W3C), 29, 85, 130

Y

Yahoo!, 12

Z

Zero-knowledge proof(s) (ZKP), 32